MOLECULES IN THE ATMOSPHERES OF EXTRASOLAR PLANETS

COVER ILLUSTRATION:

The transiting hot Jupiter HD189733b.

Credits: ESA - C.Carreau, adapted by J-P. Beaulieu.

ASTRONOMICAL SOCIETY OF THE PACIFIC CONFERENCE SERIES

A SERIES OF BOOKS ON RECENT DEVELOPMENTS IN ASTRONOMY AND ASTROPHYSICS

Volume 450

EDITORIAL STAFF

Managing Editor: Joseph Jensen
Associate Managing Editor: Jonathan Barnes
Publication Manager: Pepita Ridgeway
Editorial Assistant: Cindy Moody
e-Book Specialist: Bret Little
Web Developer/Technical Consultant: Jared M. Bellows
LATEX Consultant: T. J. Mahoney

MS 179, Utah Valley University, 800 W. University Parkway, Orem, Utah 84058-5999
Phone: 801-863-8804   E-mail: aspcs@aspbooks.org
E-book site: http://www.aspbooks.org

PUBLICATION COMMITTEE

Lynne Hillenbrand, Chair
California Institute of Technology

Marsha J. Bishop
National Radio Astronomy Observatory

Daniela Calzetti
University of Massachusetts

Gary J. Ferland
University of Kentucky

Ed Guinan
Villanova University

Luis Ho
The Observatories of the Carnegie
Institution of Washington

Scott J. Kenyon
Smithsonian Astrophysical Observatory

Doug Leonard
San Diego State University

Don McCarthy
The University of Arizona

René Racine
Université de Montréal

Ata Sarajedini
University of Florida

ASPCS volumes may be found online with color images at http://www.aspbooks.org.
ASP monographs may be found online at http://www.aspmonographs.org.

For a complete list of ASPCS volumes, ASP monographs, and
other ASP publications see http://www.astrosociety.org/pubs.html.

All book order and subscription inquiries should be directed to the ASP at
800-335-2626 (toll-free within the USA) or 415-337-2126,
or email service@astrosociety.org

ASTRONOMICAL SOCIETY OF THE PACIFIC
CONFERENCE SERIES

Volume 450

MOLECULES IN THE ATMOSPHERES OF EXTRASOLAR PLANETS

Proceedings of a conference held at
Observatoire de Paris, Paris, France
19–21 November 2008

Edited by

Jean Philippe Beaulieu
UPMC Univ Paris 06, UMR7095, Institut d'Astrophysique de Paris, F-75014, Paris, France

Stefan Dieters
School Maths and Physics, University Tasmania, Hobart, Tasmania, Australia

Giovanna Tinetti
Department Physics and Astronomy, University College London, London, UK

SAN FRANCISCO

ASTRONOMICAL SOCIETY OF THE PACIFIC
390 Ashton Avenue
San Francisco, California, 94112-1722, USA

Phone: 415-337-1100
Fax: 415-337-5205
E-mail: service@astrosociety.org
Web site: www.astrosociety.org
E-books: www.aspbooks.org

First Edition
© 2011 by Astronomical Society of the Pacific
ASP Conference Series
All rights reserved.

No part of the material protected by this copyright notice may be reproduced or utilized in any form or by any means—graphic, electronic, or mechanical, including photocopying, taping, recording, or by any information storage and retrieval system—without written permission from the Astronomical Society of the Pacific.

ISBN: 978-1-58381-782-7
e-book ISBN: 978-1-58381-783-4

Library of Congress (LOC) Cataloging in Publication (CIP) Data:
Main entry under title
Library of Congress Control Number (LCCN): 2011940728

Printed in the United States of America by Sheridan Books, Ann Arbor, Michigan.
This book is printed on acid-free paper.

Contents

Preface viii

Foreword xi
    Pierre Drossart

Participants xiv

Conference Photograph xx

Part I. Planets in the Solar System: Overview
(Chair : Athéna Coustenis)

The Composition of Planetary Atmospheres: An Historical Perspective 3
    E. Lellouch

The Molecular Ion H_3^+ in Emission in Planetary Atmospheres 19
    J. P. Maillard and S. Miller

Part II. Exoplanet Atmospheres from Transit Observations
(Chairs : Gautam Vasisht and Giovanna Tinetti)

Lessons Learned from Ground-based Transmission Spectroscopy of Extrasolar
Planets 39
    I. A. G. Snellen, S. Albrecht, E. J. W. de Mooij, and R. S. Le Poole

The Search for Exomoons 47
    D. M. Kipping

Ground Based Imaging Spectroscopy of Transiting Extrasolar Planets 55
    D. Angerhausen and A. Krabbe

Ground-based Detection of the Secondary Eclipse of TrES-3b 59
    E. J. W. de Mooij and I. A. G. Snellen

Exoplanet Spectroscopy: The Hubble Case 63
    P. Deroo, M. Swain, G. Vasisht, P. Chen, G. Tinetti, J. Bouwman,
    D. Angerhausen, and Y. Yung

Part III. Exoplanet Atmospheric Dynamics
(Chairs: Caroline Terquem and Adam Showman)

Atmospheric Dynamics of Two Eccentric Transiting Planets:
GJ 436b and HD 17156b 71
    N. K. Lewis, A. P. Showman, J. J. Fortney, M. S. Marley, and R. S. Freedman

Part IV. Molecular Data-lists and Modelling
(Chair: Bruno Bézard)

The Acetylene Laboratory IR Spectrum: New Quantitative Studies 83
    D. Jaccquemart, N. Lacome, L. Gomez, and J.-Y. Mandin (Presenter)

Collisional Line Profiles of Na Perturbed by H_2 87
    N. F. Allard

VUV Photophysics of Prebiotic Molecules in the Context of the Search for Life
on exoplanets 91
    S. Leach

Signatures of Water Clouds on Exoplanets: Numerical Simulations 101
    T. Karalidi, D. M. Stam, and C. U. Keller

On the Protoplanetary-disk Origin of the Atmospheres of Hot Super-Earths . . . 105
    M. Ikoma

Part V. Brown Dwarfs
(Chair: Jean-Pierre Maillard)

The Brown Dwarf-Exoplanet Connection 113
    A. J. Burgasser

Radiation Hydrodynamics Simulations of Dust Clouds in the Atmospheres of
Substellar Objects 125
    B. Freytag, F. Allard, D. Homeier, H-G. Ludwig, and M. Steffen

Part VI. Terrestrial Exoplanets: Modelling, Habitability, and Detection of Biosignatures (Chair: Ignasi Ribas)

The Loss of Nitrogen-rich Atmospheres from Earth-like Exoplanets within
M-star Habitable Zones 139
    H. Lammer, H. I. M. Lichtenegger, M. L. Khodachenko, Yu. N. Kulikov, and
    J.-M. Grießmeier

EXOFIT: Bayesian Estimation of Orbital Parameters of Exo-planets 147
    S. Thaithara-Balan and O. Lahav

Molecules Corresponding to a Volcanic Activity in Atmospheres of Super-Ios
and Hyper-Ios 155
Danielle Briot

Part VII. The Future: Short and Long Term Missions and Instruments to Characterise Exoplanet Atmospheres. (Chair : Vincent Coude du Foresto)

Direct Imaging of Extrasolar Planets: Overview of Ground and Space Programs . 163
A. Boccaletti

The Potential of High Contrast Coronagraphy 173
E. Serabyn

SPICA Coronagraph for the Direct Observation of Exo-Planets 181
K. Enya, and the SPICA working group

Follow the Dust: Discovery of an Exosolar Planet in Fomalhaut's Debris Disk . . 191
M. Clampin, P. Kalas, J. Graham, and E. Chiang

High-Contrast Imaging: A Wider View on Extrasolar Planetary Systems 199
M. Bonavita, R. U. Claudi, G. Tinetti, J.-L. Beuzit, G. Chauvin, S. Desidera, R. Gratton, and M. Kasper

High Contrast Imaging: A New Frontier for Exoplanets Search
and Characterization 203
R. U. Claudi, M. Bonavita, S. Desidera, R. Gratton, G. Tinetti, J.-L. Beuzit, M. Kasper, and C. Mordassini

Detailed Spectroscopy of Exoplanets Using the New Worlds Observer 209
Webster Cash and the New Worlds Observer Team

Spectral Analysis of Atmospheres by Nulling Interferometry 219
M. Ollivier and S. Jacquinod

A Spectroscopic Method for Direct Detection of Exoplanets 229
P. Cubillos, P. Rojo, and J. Fortney

Detection of Extrasolar Comets 233
O. R. Hainaut

Part VIII. The Reception

The Reception 239

Preface

The science of extra-solar planets is one of the most rapidly changing and exciting areas of astrophysics. A combination of ground-based surveys and dedicated space missions has resulted in ~700 planets being detected, and over one thousand that await confirmation. Since 1995, the number of planets known has increased by two orders of magnitude. NASAâĂŹs Kepler mission has opened up the possibility of discovering Earth-like planets in the habitable zone around some of the 100,000 stars it is surveying during its 3 to 4-year lifetime. The new ESA-Gaia mission is expected to discover thousands of new planets.

The observation of the exoplanet atmospheres is now right at the cutting edge of exoplanet science: we are making the first attempts at characering their chemical composition and temperature profiles. The ability to detect planetary atmospheric features which have $\sim 10^{-4}$ contrast with the radiation coming from its host star is quite a challenge. However, for exoplanets whose orbits are aligned so they cross the surface of their mother star when viewed from Earth, this has indeed proved to be feasible using Hubble, Spitzer and ground-based facilities. This can be done by measuring the dip in the stellar light-curve when the planet transits in front of the star (or disappears behind it), and repeating the measurement at different wavelengths. Thanks to these observations, molecules such as water vapour, methane, carbon monoxide and dioxide have been discovered for the first time in the atmosphere of an exoplanet. Whilst we started with a handful of hot giant, gaseous planets, planets such as GJ 1214b, are now within reach with current telescopes. This planet appears to be somewhat in between a rocky planet and a gaseous one, with a temperature of boiling water.

Most recently the first spectrum of a hot giant planet at a projected separation of 38 AU from its host star was observed from the ground with VLT/NACO. Spectroscopy in the shorter wavelength range of YJHK-band will likely start soon with dedicated integral field units on VLT (SPHERE) and Gemini (GPI). The young exoplanets that these instruments are expected to find and characterise are likely to feature several molecular species.

The conference was organised when the community began the characterisation of the atmospheres of exoplanets, the first direct images, and anticipated many of the next steps required. These steps are beginning to unfold now, and will continue to do so as new missions/instruments are developed. In the upcoming decade important new facilities are planned to come on line (JWST, E-ELT, TMT, SPICA), dedicated mission to characterise spectroscopically these alien worlds are being considered by NASA and ESA (FINESSE, EChO).

The main purpose of the conference was to convene different scientific communities: Solar system planetary scientists, brown dwarf and exoplanet observers and theoreticians, molecular spectroscopists and instrument experts. We covered different topics: radiative transfer, line lists, photochemical models, atmospheric dynamics, ground and space-based observations, next generation of exoplanet mission concepts from ground and space. The exoplanet revolution has indeed started and it has changed irreversibly our views about planets and stars, and the ways in which they are formed. This conference has marked a point of no return.

Jean-Philippe Beaulieu, Stefan Dieters and Giovanna Tinetti.

Scientific Organising Committee : J.P. Beaulieu, S. Dieters, T. Guillot, H. Lammer, D. Latham, D. Lin, J.P. Maillard, I. Ribas, J. Schneider, F. Selsis, J. Tennyson, G. Tinetti and S. Udry.

Local Organising Committee : V. Batista, C. Coutures, D. Kipping, D. Kubas, J.B. Marquette, C. Massin and S. Sebban.

Acknowledgments :

We are very grateful to Prof Daniel Egret, Président of the Observatoire de Paris for having hosted the conference in the Cassini Hall of Observatoire de Paris. Such a superb venue was ideal to mark the first conference fully dedicated to molecules in the atmospheres of extrasolar planets.

Special thanks to Danielle Briot and Monique Gros for their guided tour of the old Observatoire de Paris before the wine tasting session. We are very grateful to Francois Sevre for having helped us to use the 1854 "Arago Refractor" to watch the moon and Jupiter during the reception.

We acknowledge the support of the Programme "Origine des Planètes et de la Vie" and "programme National de Planetologie", from the team "Physique Stellaire et Planétaire" of the Institut d'Astrophysique de Paris and of the HOLMES project (funded by the "Agence Nationale de la Recherche"). Chrystelle Massin and Sylvie Sebban provided strong support in the preparation weeks and during the very dense three days of the meeting.

Foreword

Entering a new era : the characterization of exoplanets

After fifteen years mostly devoted to the detection of new planets, exoplanetary science has only recently entered a new domain in the physics of these extraordinary objects, with the interpretation of the first detailed spectra of exoplanets. Making a comparison with the history of planetary science, we are today in the situation of the first observations of methane and ammonia on the giant planets (Kuiper, 1947), performed in the 30s and 40s years of the last century: poor resolution, controversial results and hot discussions on the first interpretations, but the data are here and are improving year after year. We must remember these papers in the 50s years on the spectroscopic detection of vegetation on Mars (Sinton 1957), not to blame the authors, but to know that errors and corrections are inherent to the scientific method, and also to avoid to repeat the same errors today. The conference "Molecules in the atmospheres of extrasolar planets" provides a complete overview of the state of the art in terms of molecular spectroscopy adapted to Solar System planets, extrasolar planets and brown dwarf research.

An overview of the planetary research is the obvious starting point of the Conference. After 50 years in the space age, and the exploration of a large part of the Solar System with various instrumentation, the knowledge of the objects themselves, and of the physical processes sculpting their spectra gives today many important guidelines for extrasolar planets studies. First of all is the remark that even very minor constituents, in the range of a ppm or lower can have important effects on the spectroscopy of planets, like PH_3 on Jupiter and Saturn at $5\mu m$. This is due to a combination of physical processes, like highly inhomogeneous atmospheric structure, the presence of out of equilibrium chemistry and also to the complexity of radiative transfer equations, even within the range of local thermal equilibrium. One aspect that Earth and planetary scientists have developed during decennials is the difficulty to simplify the radiative transfer equation into more tractable "band models"; the smooth shape of low resolution spectra of planets masks the high complexity of high resolution spectra, originating from different atmospheric levels. Accurate modern modelling is today benchmarked to line-by-line calculations in radiative transfer equations, which for many molecular bands may mean to handle several hundred thousands molecular lines.

Therefore, up to date database of molecular bands is an important part of the modelling in extrasolar planets, and Part IV in the Conference was devoted to some aspects of this problem. The infrared spectral range is the most adapted to planetary molecular spectroscopy, most of the vibration-rotation fundamental bands falling in this range (Herzberg 1945). The interpretation of molecular spectra of even simple molecules proves to be incredibly complex, due to a variety of quantum effects, only part of them being tractable. A naive approach could be that for low resolution spectra, with simple band detections, a simplified approach could be sufficient. The experience from Earth and planetary studies shows that this is an illusion, because convolution at low resolution mixes emission from different layers, and an intuitive approach, which may be valid in the interpretation of atomic lines in stellar atmospheres, fails in planetary atmospheres and molecular spectra. One of the complexity, difficult to handle in spectral modelling even for a well known planet like Earth, is the cloud contribution, coupled with composition variation but with only weak spectroscopic signatures.

At the heart of the Conference is the new and fascinating spectral observations of transits and occultations by exoplanets (Part II). Today, only a handful of objects allows the investigators to retrieve relevant spectral information from the absorption or the emission of their atmospheres. As usual in new and exploratory science, many results are still controversial, and firmly confirmed evidences are often lacking, but is is clear that in the coming years, more results will arrive and the comparison with models will improve. The results presented here therefore mark an important step in the development in this new science, from the confirmed evidence for atomic lines like Na in exoplanets, to the molecular spectra progressively emerging from low signal to noise and coarse spectrophometric observations.

The discussion on the importance of dynamics effects (Part III) is certainly a far reaching objective, for objects whose composition and structure is only poorly constrained, but this question is of high importance in Solar System atmospheres. "Our" planets have a large variety of dynamical effects, from the superrotation of Venus to differential rotation of giant planets, which cannot be neglected even at first order, if we want to understand the outer aspects of exoplanets.

If the comparison between exoplanets and solar system planets is obviously important, the comparison with brown dwarfs (Part V) is at the other extreme of equal importance. Firstly, brown dwarfs are relatively simpler to study than exoplanets and their spectral exploration is today at a higher level of understanding. The evidence for the importance of cloud effects, non equilibrium chemistry and inhomogeneities in their atmosphere will have to be translated in the conditions of the exoplanets. In addition, the question of exoplanet-brown dwarf connection addressed in this part is a fascinating subject connected to general astrophysical questions of the formation of our Universe.

Terrestrial planets are the far reaching goal of exoplanetary research, and the Conference reports the status of modelling and detection of such objects (Part VI). The difficulty in retrieving information there is due not only to their intrinsic faintness, but also to more compact atmospheres if we refer to our known Solar System. A rough definition, from a spectroscopic point of view, which is valid in the Solar system, will refer to terrestrial planets as CO_2 rich planets, in opposition with giant planets as CH_4 spectrally dominated planets. Whether or not this distinction remains valid for exoplanets is an important question that will probably be addressed in a near future, especially for the new class of objects called "Superearths" that are today detected.

The Conference closes with future perspectives, extrapolating from the present status of observations and modelling of exoplanet atmospheres. In the exploration of the new worlds we are pioneering today, there is only one certainty: we will be surprised! Many of today's a priori will have to be revisited in order to improve our knowledge of the Universe. Coming back to our Solar System, planetary research is already boosted by some questions raised in the exoplanet research: the diversity of spectra will extend our planetary models to new domains. The origin of the solar system, with the question of planetary migration, has to be reworked. Concerning the Earth itself, we can expect from exoplanets observations to learn more about the habitability, the evolution of atmospheres (e.g. :are Venus-like atmospheres the most common state of telluric planets ?). We are entering a new era.

Pierre Drossart
LESIA, Observatoire de Paris, CNRS, Université Pierre et Marie Curie, Université Paris-Diderot.

Kuiper G. P. Infrared Spectra of Planets. Astrophysical Journal, vol. 106, p.251 (1947)
Sinton W. M. Spectroscopic Evidence for Vegetation on Mars. Astrophysical Journal, vol. 126, p.231 (1957)
Herzberg, G. Molecular spectra and molecular structure. Vol.2: Infrared and Raman spectra of polyatomic molecules.New York: Van Nostrand, Reinhold, 1945

Participants

OLIVIER ABSIL, Laboratoire d'Astronomie de l'Observatoire de Grenoble, 414 rue de la Piscine, F-38041, Grenoble, France ⟨ olivier.absil@obs.ujf-grenoble.fr ⟩

FRANCE ALLARD, Ecole Normale Supérieure de Lyon, 46 Allée d'Italie, 69007 Lyon, France ⟨ fallard@ens-lyon.fr ⟩

NICOLE ALLARD, LERMA, Observatoire de Paris, 77 Avenue Denfer Rochereau, 75104 Paris, France ⟨ nicole.allard@obspm.fr ⟩

DAVID ANDERSON, Astrophysics Group, Keele University, Staffordshire, ST5 5BG, United Kingdom ⟨ dra@astro.keele.ac.uk ⟩

DANIEL ANGERHAUSEN, Jet Propulsion Laboratory, M/S 301-451, 4800 Oak Grove Dr., Pasadena, CA 91109, USA ⟨ anger@ph1.uni-koeln.de ⟩

ALAN AYLWARD, University College London, Gower Street, London, WC1E 6BT, United Kingdom ⟨ a.aylward@ucl.ac.uk ⟩

MAURO BARBIERI, CISAS c/o Dept of Physics, via Marzolo 8, 35131 Padova, Italy ⟨ mbarbier@pd.infn.it ⟩

JOHN BARNES, University of Hertfordshire, Centre for Astrophysics Research, College Lane, Hatfield AL10, 9AB, United Kingdom ⟨ j.r.barnes@herts.ac.uk ⟩

VIRGINIE BATISTA, Institut d'Astrophysique de Paris, 98bis Boulevard Arago, 75014 Paris, France ⟨ batista@iap.fr ⟩

PIERRE BAUDOZ, LESIA, Observatoire de Paris-Meudon, 5 Place Jules Janssen 92190 Meudon, France ⟨ pierre.baudoz@obspm.fr ⟩

JEAN-PHILIPPE BEAULIEU, Institut d'Astrophysique de Paris, 98bis Boulevard Arago, 75014 Paris, France ⟨ beaulieu@iap.fr ⟩

CHAD BENDER, Dept. of Astronomy and Astrophysics, The Pennsylvania State University, 525 Davey Lab University Park, PA 16802, USA ⟨ cbender@psu.edu ⟩

DANIEL BENEST, Observatoire de la Côte d'Azur UMR 6202 Cassiopée / CNRS, B.P. 4229 06304 Nice Cedex 4, France ⟨ benest@oca.eu ⟩

SLIMANE BENSAMMAR, Observatoire de Paris, 77 Avenue Denfer Rochereau, 75104 Paris, France ⟨ slimane.bensammar@obspm.fr ⟩

JEAN-LUC BEUZIT, Laboratoire d'Astronomie de l'Observatoire de Grenoble, 414 rue de la Piscine, F-38041, Grenoble, France ⟨ jean-luc.beuzit@obs.ujf-grenoble.fr ⟩

BRUNO BEZARD, LESIA, Observatoire de Paris-Meudon, 5 Place Jules Janssen 92190 Meudon, France ⟨ Bruno.Bezard@obspm.fr ⟩

ANTHONY BOCCALETTI, LESIA, Observatoire de Paris-Meudon, 5 Place Jules Janssen 92190 Meudon, France ⟨ anthony.boccaletti@obspm.fr ⟩

MARIANGELA BONAVITA, Osservatorio Astronomico di Padova - INAF, Vicolo dell'Osservatorio 5, I - 35122 Padova, Italy ⟨ mariangela.bonavita@oapd.inaf.it ⟩

MICKAEL BONNEFOY, Laboratoire d'Astronomie de l'Observatoire de Grenoble, 414 rue de la Piscine, F-38041, Grenoble, France ⟨ mbonnefo@obs.ujf-grenoble.fr ⟩

DANIELLE BRIOT, Observatoire de Paris, 77 Avenue Denfer Rochereau, 75104 Paris, France ⟨ danielle.briot@obspm.fr ⟩

ADAM BURGASSER, University of California San Diego Center for Astrophysics and Space Science, 9500 Gilman Drive, Mail Code 0424 La Jolla, CA 92093, USA ⟨ aburgasser@ucsd.edu ⟩

ANDREW CARTER, The Open University,, PO Box 197 Milton Keynes, MK7 6BJ United Kingdom ⟨ a.carter@open.ac.uk ⟩

JOSH CARTER, Harvard-Smithsonian Center for Astrophysics, 60 Garden St., Cambridge MA 02138, USA ⟨ jacarter@cfa.harvard.edu ⟩

WEBSTER CASH, Department of Astrophysical and Planetary Science, Center for Astrophysics and Space Astronomy University of Colorado, Campus Box 389 Boulder, CO 80309-0389 ⟨ webster.cash@colorado.edu ⟩

DAVID CHARBONNEAU, Harvard-Smithsonian Center for Astrophysics, 60 Garden St., Cambridge MA 02138, USA ⟨ dcharbonneau@cfa.harvard.edu ⟩

PIN CHEN, Jet Propulsion Laboratory, 4800 Oak Grove Dr., Pasadena, CA 91109, USA ⟨ pin.chen@jpl.nasa.gov ⟩

JAMES CHO, School of Mathematical Sciences Queen Mary, University of London, Mile End Road London E1 4NS, United Kingdom ⟨ J.Cho@qmul.ac.uk ⟩

MARK CLAMPIN, Astrophysics Science Division, NASA/GSFC, Code 667, Exoplanets and Stellar Astrophysics Laboratory Greenbelt, MD 20771, USA ⟨ mark.clampin@nasa.gov ⟩

RICCARDO CLAUDI, Osservatorio Astronomico di Padova - INAF, Vicolo dell'Osservatorio 5, I - 35122 Padova, Italy ⟨ riccardo.claudi@oapd.inaf.it ⟩

VINCENT COUDE DU FORESTO, LESIA, Observatoire de Paris-Meudon, 5 Place Jules Janssen 92190 Meudon, France ⟨ vincent.foresto@obspm.fr ⟩

CHRISTIAN COUTURES, Institut d'Astrophysique de Paris, 98bis Boulevard Arago, 75014 Paris, France ⟨ coutures@iap.fr ⟩

ERNST DE MOOIJ, Leiden Observatory, Huygens Laboratory J.H. Oort Building, P.O. Box 9513, NL-2300 RA Leiden, The Netherlands ⟨ demooij@strw.leidenuniv.nl ⟩

PATRICIO CUBILLOS, Planetary Sciences Group, Department of Physics, University of Central FloridaOrlando, FL 32816-2385, USA ⟨ pcubillos@das.uchile.cl ⟩

KARLA DE SOUZA TORRES, Universidade Tecnológica Federal do Parana - UTFPR Departamento de Informatica, DAINF, Av. Sete de Setembro, 3165, Reboucas Curitiba-PR, CEP 80230-901, Brazil ⟨ karlchen79@gmail.com ⟩

PIETER DEROO, Jet Propulsion Laboratory, California Institute of Technology, 4800 Oak Grove Drive, Pasadena, California 91109-8099, USA ⟨ pieter.d.deroo@jpl.nasa.gov ⟩

JEAN-MICHEL DESERT, Institut d'Astrophysique de Paris, 98bis Boulevard Arago, 75014 PARIS, France ⟨ desert@iap.fr ⟩

Participants

RENE DOYON, Université de Montréal, Département de physique, Université de Montréal, C.P. 6128, succ. centre-ville, Montréal (Québec) H3C 3J7, Canada ⟨doyon@astro.umontreal.ca⟩

PIERRE DROSSART, LESIA, Observatoire de Paris-Meudon, 5 Place Jules Janssen 92190 Meudon, France ⟨pierre.drossart@obspm.fr⟩

DANIEL EGRET, LESIA, Observatoire de Paris-Meudon, 5 Place Jules Janssen 92190 Meudon, France ⟨daniel.egret@obspm.fr⟩

THERESE ENCRENAZ, LESIA, Observatoire de Paris-Meudon, 5 Place Jules Janssen 92190 Meudon, France ⟨therese.encrenaz@obspm.fr⟩

KEIGO ENYA, Department of Infrared Astrophysics, Institute of Space and Astronautical Science, Japan Aerospace Exploration Agency Yoshinodai 3-1-1, Sagamihara, Kanagawa 229-8510 ⟨enya@ir.isas.jaxa.jp⟩

NICOLE FEAUTRIER, LERMA, Observatoire de Paris-Meudon, 5 Place Jules Janssen 92190 Meudon, France ⟨Nicole.Feautrier@obspm.fr⟩

RICHARD FREEDMAN, SETI Institute, NASA Ames Research Center, MS 254-5, Moffett Field, CA 94035, USA ⟨freedman@darkstar.arc.nasa.gov⟩

BERND FREYTAG, Ecole Normale Supérieure de Lyon, 46 Allée d'Italie, 69007 Lyon, France ⟨Bernd.Freytag@ens-lyon.fr⟩

MARYVONNE GERIN, LERMA, ENS, 24 r Lhomond FR 75231 Paris Cedex 05. France ⟨maryvonne.gerin@ens.fr⟩

LAURA GOMEZ-MARTIN, Université Pierre et. Marie Curie. LADIR., 4 place Jussieu, 75005, Paris, France ⟨lauragm78@hotmail.com⟩

IOULI GORDON, Harvard-Smithsonian Center for Astrophysics, 60 Garden St., Cambridge MA 02138, USA ⟨igordon@cfa.harvard.edu⟩

JOHN LEE GRENFELL, Zentrum für Astronomie und Astrophysik (ZAA), Technische Universität Berlin (TUB) Hardenbergstr. 36 10623 Berlin Germany ⟨lee.grenfell@dlr.de⟩

TRISTAN GUILLOT, Observatoire de la Cote d'Azur CNRS / Laboratoire Cassiopée, B.P. 4229, 06304 Nice Cedex 4, France ⟨guillot@obs-nice.fr⟩

OLIVIER HAINAUT, European Southern Observatory, Karl-Schwarzschild-Str. 2. D-85748 Garching bei München Germany ⟨ohainaut@eso.org⟩

JOSEPH HARRINGTON, Planetary Sciences Group, Department of Physics, University of Central FloridaOrlando, FL 32816-2385, USA ⟨jh@physics.ucf.edu⟩

PASCAL HEDELT, Zentrum für Astronomie und Astrophysik (ZAA), Technische Universität Berlin (TUB) Hardenbergstr. 36 10623 Berlin Germany ⟨pascal.hedelt@dlr.de⟩

MAJDI HOCHLAF, B. N34, Bat Lavoisier Laboratoire MSME, U. Paris-Est Marne-La-Vallée, 5 Bd Descartes, 77454 Champs sur Marne, France ⟨hochlaf@univ-mlv.fr⟩

DEREK HOMEIER, Centre de Recherche Astrophysique de Lyon, ENS Lyon 46 allée d'Italie 69364 Lyon Cedex 07 ⟨dhomeie@gwdg.de⟩

MASAHIRO IKOMA, Ida Laboratory, Earth and Planetary Sciences, Tokyo Institute of Technology, 1-2, 2-11-1, Ookayama, Meguro-ku, Tokyo, 152-8851, Japan ⟨ mikoma@geo.titech.ac.jp ⟩

NICOLAS IRO, Lennard-Jones Laboratories, Keele University, Staffordshire ST5 5BG, United Kingdom ⟨ iro@astro.keele.ac.uk ⟩

DAVID JACQUEMART, Université Pierre et. Marie Curie. LADIR., 4 place Jussieu, 75005, Paris, France ⟨ david.jacquemart@upmc.fr ⟩

THEODORA KARALIDI, SRON Netherlands Institute for Space Research, Sorbonnelaan 2. NL 3584 CA Utrecht, The Netherlands ⟨ androm3da1@gmail.com ⟩

DAVID KIPPING, Harvard-Smithsonian Center for Astrophysics, 60 Garden St., Cambridge MA 02138, USA ⟨ dkipping@cfa.harvard.edu ⟩

HEATHER KNUTSON, California Institute of Technology, Division of Geological & Planetary Sciences, 1200 E California Blvd MC 150-21, Pasadena, CA 91125 USA ⟨ hknutson@caltech.edu ⟩

TOMMI KOSKINEN, Department of Planetary Sciences, Space Science Bldg, Room 525, 1629 E. University Blvd Tucson, AZ 85721-0092, USA ⟨ tommi@lpl.arizona.edu ⟩

TAKAYUKI KOTANI, Department of Infrared Astrophysics, Institute of Space and Astronautical Science, Japan Aerospace Exploration Agency (JAXA), 3-1-1 Yoshinodai, Chuo-ku, Sagamihara 252-5210 ⟨ kotani@ir.isas.jaxa.jp ⟩

DANIEL KUBAS, Institut d'Astrophysique de Paris, 98bis Boulevard Arago, 75014 PARIS, France ⟨ dkubas@gmail.com ⟩

OFER LAHAV, University College London, Gower Street, London, WC1E 6BT, United Kingdom ⟨ lahav@star.ucl.ac.uk ⟩

HELMUT LAMMER, Space Research Institute, Austrian, Academy of Sciences, Schmiedlstr. 6, A-8042 Graz, Austria ⟨ helmut.lammer@oeaw.ac.at ⟩

SYDNEY LEACH, LERMA, Observatoire de Paris-Meudon, 5 Place Jules Janssen 92190 Meudon, France ⟨ Sydney.Leach@obspm.fr ⟩

EMMANUEL LELLOUCH, LESIA, Observatoire de Paris-Meudon, 5 Place Jules Janssen 92190 Meudon, France ⟨ emmanuel.lellouch@obspm.fr ⟩

NIKOLE LEWIS, Department of Planetary Sciences, Space Science Bldg, Room 525, 1629 E. University Blvd Tucson, AZ 85721-0092, USA ⟨ nlewis@lpl.arizona.edu ⟩

NIKKU MADHUSUDHAN, Department Astrophysical Sciences,, Princeton University, Princeton, NJ 08544, USA ⟨ nmadhu@astro.princeton.edu ⟩

JEAN-PIERRE MAILLARD, Institut d'Astrophysique de Paris, 98bis Boulevard Arago, 75014 Paris, France ⟨ maillard@iap.fr ⟩

JEAN-BAPTISTE MARQUETTE, Institut d'Astrophysique de Paris, 98bis Boulevard Arago, 75014 Paris, France ⟨ marquett@iap.fr ⟩

EDUARDO MARTIN, Centro de Astrobiología (INTA-CSIC), Carretera de Ajalvir km 4, Torrejón de Ardoz, 28850 Madrid, Spain ⟨ ege@iac.es ⟩

CLAIRE MARTIN-ZAIDI, Laboratoire d'Astronomie de l'Observatoire de Grenoble, 414 rue de la Piscine, F-38041, Grenoble, France ⟨ claire.martin-zaidi@obs.ujf-grenoble.fr ⟩

FRANCOIS MENARD, Laboratoire d'Astronomie de l'Observatoire de Grenoble, 414 rue de la Piscine, F-38041, Grenoble, France ⟨ menard@obs.ujf-grenoble.fr ⟩

BERTRAND MENESSON, Jet Propulsion Laboratory, 4800 Oak Grove Dr., Pasadena, CA 91109, USA ⟨ bertrand.menesson@jpl.nasa.gov ⟩

STEVE MILLER, University College London, Gower Street, London, WC1E 6BT, United Kingdom ⟨ s.miller@ucl.ac.uk ⟩

RAPHAEL MORENO, LESIA, Observatoire de Paris-Meudon, 5 Place Jules Janssen 92190 Meudon, France ⟨ raphael.moreno@obspm.fr ⟩

NORIO NARITA, National Astronomical Observatory of Japan, 2-21-1 Osawa, Mitaka, Tokyo, 181-8588, Japan ⟨ norio.narita@nao.ac.jp ⟩

MARC OLLIVIER, Institut d'Astrophysique Spatiale, Bâtiment 121, Université Paris XI 91405 Orsay cedex - France ⟨ marc.ollivier@ias.u-psud.fr ⟩

ENRIC PALLE, Instituto de Astrofísica de Canarias (IAC), Vía Láctea s/n 38200, La Laguna, Tenerife, Spain ⟨ epalle@iac.es ⟩

IGNASI RIBAS, IEEC-CSIC, Campus UAB, Torre C5 - Parell - 2a planta, 08193 Bellaterra, Spain ⟨ iribas@ieec.uab.es ⟩

FRANCOIS ROBERT, Laboratoire de Minéralogie et Cosmochimie du Muséum (LMCM), UMR 7202 CNRS, Case 52 - 57 Rue Cuvier / 61 rue Buffon - 75231 Paris Cedex 05 ⟨ robert@mnhn.fr ⟩

FLORIAN RODLER, IEEC-CSIC, Campus UAB, Torre C5 - Parell - 2a planta, 08193 Bellaterra, Spain ⟨ f.rodler@yahoo.com ⟩

LAURENCE ROTHMAN, Harvard-Smithsonian Center for Astrophysics, 60 Garden St., Cambridge MA 02138, USA ⟨ LRothman@CfA.Harvard.edu ⟩

EVELYNE ROUEFF, LERMA, Observatoire de Paris-Meudon, 5 Place Jules Janssen 92190 Meudon, France ⟨ evelyne.roueff@obspm.fr ⟩

JEAN SCHNEIDER, LUTH, Observatoire de Paris-Meudon, 5 Place Jules Janssen 92190 Meudon, France ⟨ jean.schneider@obspm.fr ⟩

FRANCK SELSIS, Laboratoire d'Astrophysique de Bordeaux,, (Université Bordeaux 1), B.P. 89, 33271 Floirac Cedex, France ⟨ selsis@obs.u-bordeaux1.fr ⟩

EUGENE SERABYN, Jet Propulsion Laboratory, 4800 Oak Grove Dr., Pasadena, CA 91109, USA ⟨ gene.serabyn@jpl.nasa.gov ⟩

ADAM SHOWMAN, Department of Planetary Sciences and Lunar and Planetary Laboratory, The University of Arizona, 1629 University Blvd., Tucson, AZ 85721, USA ⟨ showman@lpl.arizona.edu ⟩

IAN SIMS, Institut de Physique de Rennes, Equipe : "Astrochimie Expérimentale", UMR CNRS 6251, Université de Rennes 1, Campus de Beaulieu, 35042, Rennes Cedex, France ⟨ ian.sims@univ-rennes1.fr ⟩

DAVID SING, Astrophysics Group, School of Physics, University of Exeter, Stocker Road, Exeter EX4 4QL, United Kingdom ⟨ sing@astro.ex.ac.uk ⟩

IGNAS SNELLEN, Leiden Observatory, Leiden University, Postbus 9513, NL-2300 RA Leiden, The Netherlands ⟨ snellen@strw.leidenuniv.nl ⟩

ANNIE SPIELFIEDEL, LERMA, Observatoire de Paris-Meudon, 92195, Meudon Cedex, France ⟨ annie.spielfiedel@obspm.fr ⟩

DAPHNE STAM, SRON Netherlands Institute for Space Research, Sorbonnelaan 2. NL 3584 CA Utrecht, The Netherlands ⟨ daphne@sron.nl ⟩

MARK SWAIN, Jet Propulsion Laboratory, 4800 Oak Grove Dr., Pasadena, CA 91109, USA ⟨ mark.r.swain@jpl.nasa.gov ⟩

BRUCE SWINYARD, University College London, Gower Street, London, WC1E 6BT, United Kingdom ⟨ bruce.swinyard@stfc.ac.uk ⟩

DAHBIA TALBI, Université Montpellier II, Groupe de Recherches en Astronomie et Astrophysique du Languedoc,, CNRS, UMR 5024, place Eugene Bataillon, 34095 Montpellier, France ⟨ dahbia.talbi@graal.univ-montp2.fr ⟩

JONATHAN TENNYSON, University College London, Gower Street, London, WC1E 6BT, United Kingdom ⟨ j.tennyson@ucl.ac.uk ⟩

CAROLINE TERQUEM, Institut d'Astrophysique de Paris, 98bis Boulevard Arago, 75014 Paris, France ⟨ caroline.terquem@iap.fr ⟩

SREEKUMAR THAITHARA BALAN, Department of Physics, Cavendish Laboratory, University of Cambridge, J J Thomson Avenue, Cambridge, CB3 0HE, United Kingdom ⟨ st452@mrao.cam.ac.uk ⟩

GIOVANNA TINETTI, University College London, Gower Street, London, WC1E 6BT, United Kingdom ⟨ g.tinetti@ucl.ac.uk ⟩

WESLEY TRAUB, Jet Propulsion Laboratory, 4800 Oak Grove Dr., Pasadena, CA 91109, USA ⟨ wtraub@jpl.nasa.gov ⟩

GAUTAM VASISHT, Jet Propulsion Laboratory, 4800 Oak Grove Dr., Pasadena, CA 91109, USA ⟨ gautam.vasisht@jpl.nasa.gov ⟩

YUK YUNG, California Institute of Technology, Division of Geological & Planetary Sciences, 1200 E California Blvd MC 150-21, Pasadena, CA 91125 USA ⟨ yly@gps.caltech.edu ⟩

Attendees outside the observatoire de Paris : 1 P. Baudoz, 2 J.P. Maillard, 3 T. Karalidi, 4 A. Boccaleti, 5 M. Ikoma, 6 D. Stam, 7 J. Schneider, 8 J. Barnes, 9 O. Lahav, 10 N. Iro, 12 G. Serabyn, 13 A Coustenis, 14 C. Griffith, 15 G. Tinetti, 16 J.P. Beaulieu, 17 H. Lammer, 18 O. Absil, 19 M. Barbieri, 20 M. Clampin, 21 K. Enya, 22 F. Selsis, 24 A.S. Maurin, 25 R. Claudi, 26 M. Bonavita, 27 J.M. Desert, 28 I. Snellen, 29 D. Sing, 30 E. De Mooi, 31 C. Bender, 32 P. Deroo, 33 B. Bezard, 34 L. Rothmans, 35 L. Grenfell, 36 E. Palle, 37 T. Kotani, 39 M. Ollivier, 40 N. Narita, 41 D. Kipping, 42 V. Coude du Forresto, 43 I. Ribas, 44 T. Yamashita, 45 M. Swain, 46 W. Cash, 47 D. Angerhauser, 48 W. Traub, 49 A. Leger and 50 S. Bensamar.

Part I

Planets in the Solar System: Overview
(Chair : Athéna Coustenis)

The Composition of Planetary Atmospheres: An Historical Perspective

Emmanuel Lellouch

Observatoire de Paris, F-92195 Meudon, France

Abstract. A brief review of the composition of planetary atmospheres is given. By adopting a historical perspective, and without pretending to exhaustiveness, we attempt to show how the evolution of observational knowledge progressively gave access to insight into the main physical processes ruling planetary atmospheres, and to illustrate the synergy between observations from the ground and from space.

1. Introduction

Atmospheres in the Solar System can be classified in four broad categories. *Giant Planets* (Jupiter, Saturn, Uranus, Neptune) have massive atmospheres, composed primarily of molecular hydrogen, atomic helium, and species such as methane and ammonia. These atmospheres are termed primary, in the sense that they are contemporary to the planet formation, having been directly captured from Solar nebula gases and expelled from protoplanetary ices. Due to the large masses and cold temperatures, and to the absence of surface interaction, these atmospheres have not evolved much since planetary formation. *Terrestrial planets* (Venus, Earth, Mars, and for that purpose Titan) have secondary atmospheres, composed mainly of heavy gases (e.g. N_2, O_2, CO_2), which were outgassed subsequent to planet formation, and have strongly evolved in terms of pressure and/or major composition since formation, through atmospheric escape and surface interactions. *Tenuous atmospheres* (in the nanobar to microbar range) are present on the surface of several icy bodies with sufficient gravity, including Pluto (N_2), Triton (N_2), Io (SO_2), and to a lesser extent Enceladus (H_2O). These atmospheres reflect equilibrium with sublimating surface ices or internal (e.g. volcanic or cryovolcanic) sources. Even more tenuous atmospheres, at the level of the *exospheric limit* ($\sim 10^{-12}$ bar) are present around Europa (O_2), Ganymede (O_2), Callisto (CO_2), as well as even much more tenuous exospheres (Na-dominated) on the Moon and Mercury. Here, we review the composition of "dense" atmospheres (i.e. the first two categories), trying to illustrate as much as possible the stream of historical discoveries and ideas.

2. Early Times

Observational evidence for molecules in planets has been obtained over a century ago. In 1905, W.M. Slipher from Lowell Observatory obtained photographic spectra of Saturn and Jupiter in which he found strong absorption bands. It was not until 1932, however, that these bands were identified as methane and ammonia (Wildt 1932). Shortly later, Kuiper (1944), (Kuiper 1949) observed methane in the photographic infrared spectra of Uranus, Neptune and Titan (Fig. 1). Although Kuiper did not seem particularly interested in its discovery of methane on Titan ("The only reason why I happened to observe the planets and the ten brightest satellites was that they were nicely lined up in a region of the sky where I had run out of programs stars."), it was one of utmost importance. Indeed, methane is now recognized to be second in abundance and first in importance molecule in Titan's atmosphere, controlling both stratospheric chemistry (along with N_2) and tropospheric meteorology, and playing essentially the physical role water plays on Earth (cloud formation, rain, lakes, landscape shaping...). Visible spectra of the Giant Planets showed a few other weak bands, not due to methane eventually, shown (Herzberg 1952; Spinrad 1963a) to be due to molecular hydrogen.

Figure 1. The detection of methane in Titan (Kuiper 1944).

As for telluric planets, (Adams & Dunham 1932) reported the possible detection of CO_2 at 7820 Å in the spectrum of Venus. Unambiguous confirmation, however, had to await the development of IR detectors derived from military technology, first in the USA (e.g. Kuiper 1947)) and as early as in the 1950s in the Soviet Union., V.I. Moroz (1931-2004), the father of soviet planetary science, exploited these new PbS detectors to perform extensive studies of the IR spectra of Venus and Mars. The paper on Mars by Moroz (1964) is probably one of the first examples of a "modern-style" solar system paper, with informative figures (see Fig. 2), a well-documented abstract, and several quantitatively correct conclusions, such as *"The pressure at the planet's surface is evidently much lower than 100 millibars. The presence of ice absorption bands in the polar cap spectrum is confirmed"*.

An extraordinary claim by Sinton (1957, "Spectroscopic evidence for vegetation on Mars"), based on the presence of a small depression at 3.45 μm in the spectrum of Mars attributed to the signature of lichens, but later dismissed as due to deuterated water in the Earth's atmosphere (Rea et al. 1965), illustrated the danger of low spectral resolution – and probably as well of preconceived ideas. Yet, this epoch, and up to the early 1970s, was central for understanding the martian physico-chemistry, with the detection of Mars' main atmospheric gases: H_2O at 0.82 μm (Spinrad 1963b)—now known to be subject to a complex seasonal cycle (Fig. 3), involving the interaction with polar caps, clouds, and probably regolith reservoirs—CO in the near-IR (Kaplan et al. 1969), O_3 in the UV from the Mariner 9 mission (Barth & Hord 1971) and O_2 in the visible (Carleton & Traub 1972). These breakthrough discoveries allowed the basic

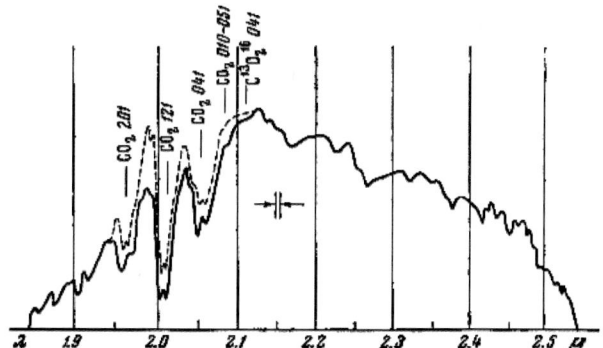

Fig. 2. Martian spectrum in 1.85-2.55 μ region. Grating spectrometer, PbS with CO₂ cooling, mean of three recordings of March 5, 1963, Z = 34-44°, slit 1.5 mm (Δλ = 60 A), rate 7.3 A/sec, τ = 4.4 sec. Solar spectrum shown by dashed line.

Figure 2. An IR spectrum of Mars from Moroz (1964), with its caption.

chemical cycles to be established, such as:

$$CO_2 + h\nu \longrightarrow CO + O \quad (1)$$

$$O + O + M \longrightarrow O_2 + M \quad (2)$$

$$O_2 + O + M \longrightarrow O_3 + M \quad (3)$$

$$H_2O + h\nu \longrightarrow OH + H \quad (4)$$

$$CO + OH \longrightarrow CO_2 + H \quad (5)$$

where the latter reaction is key in maintaining the stability of the CO_2 atmosphere in spite of its photolysis in reaction (1). The odd H radicals produced in reaction (4) can undergo further reactions, such as $OH + O + M \longrightarrow HO_2 + M$, and $HO_2 + HO_2 \longrightarrow H_2O_2 + O_2$, thereby producing hydrogen peroxide. The latter, however, was not detected until 2003 (Clancy et al. 2004; Encrenaz et al. 2004).

Similarly, near-IR observations lead to the discovery of several molecules (HCl, HF, CO) in the near-IR solar component reflected off Venus clouds (Connes et al. 1967), revealing notably the presence of halides in the Venus atmosphere. In contrast, H_2O, O_2 and O_3 were not unambiguously detected in Venus. In addition, these observations were restricted to the upper atmosphere (> 70 km) of Venus, while the deep atmosphere below the cloud tops could not be probed.

3. The 1970s: Main Concepts Emerge

The decade of the 70's was characterized by the opening and first systematic exploitation of the *thermal infrared* range. Although the first radiometric observations of the 10-μm window were actually obtained as early as 1954 (Sinton & Strong 1960), in the case of Mars, revealing for the first time the thermal behaviour of the martian surface and detecting new CO_2 bands), it was not until the early 1970s that spectro-photometric and spectroscopic techniques using new photoconductor detectors (such as CuGe, GeS...)

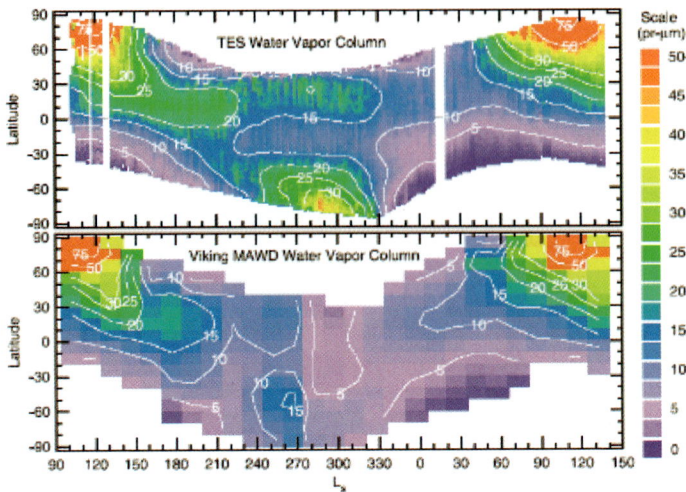

Figure 3. The water cycle on Mars, showing the column abundance of water vapor (precipitable micron) as a function of season and latitude (from Smith 2002).

led to the discovery of new species and access to new physical concepts. Indeed, by virtue of the radiative transfer equation, thermal emission is sensitive to both the thermal structure and the vertical distribution of gases in an atmosphere. This makes the thermal component much more information-rich than the solar component, which is to first order insensitive to gas temperature, and only depends on the column-integrated amount of molecules above the reflecting/scattering layer.

The first explorations of the spectrum of Jupiter, Saturn, and Titan in the 10-μm telluric transmission window were particularly interesting, revealing emission from several hydrocarbons in addition to methane, such as ethane (C_2H_6), acetylene (C_2H_2), ethylene (C_2H_4) and deuterated methane (CH_3D) (Fig. 4). This had two immediate and essential implications, outlined hereafter.

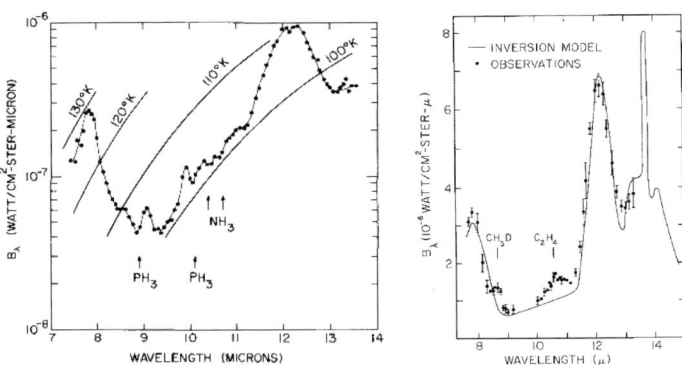

Figure 4. Some early spectra of Saturn (left) and Titan (right) in the 7-13 μm window (from Gillett & Forest (1974); Gillett (1975)).

3.1. Stratospheres of the Outer Planets

The discovery of *emission* features implied that these objects exhibited stratospheres, for which a temperature could be estimated. For example, the much more contrasted emission features of ethane on Titan than on Saturn (Fig. 4), implied a warmer stratosphere, ~170 K on Titan vs ~140 K on Saturn. The case for a warm stratosphere on Titan had in fact been anticipated, as it was known already, based on the remarkably low UV albedo of Titan (Danielson et al. 1973), that Titan's haze absorbs solar energy. Although at that time fundamental parameters of Titan's atmosphere, such as the surface pressure and the nature of the major consituent, were still unknown, these discoveries led to the development of thermal models based on physical principles and fitting the thermal IR data. Essentially, in the 1970s, two competing models were in vogue, one in which Titan's atmosphere was dominantly CH_4 and restricted to a stratosphere with ~20 mbar pressure, and one in which the atmosphere included an additional dominant gas (ammonia, nitrogen, a noble gas...) and a deep troposphere due to greenhouse effect by H_2 (detected by Trafton (1975)) and CH_4. In the latter model, the surface pressure was of several hundred of millibars or more, depending on the actual surface temperature – improperly measured at that time. Although it is that model, with N_2 as the dominant species and 94 K, 1.5 bar, as surface conditions, that turned out to be confirmed by Voyager 1 in 1981 (e.g. Lindal et al. 1983), the two classes of models had correctly captured the physics of stratospheric heating (also known as anti-greenhouse effect) by a high-altitude haze, opaque to visible light and transparent in the thermal IR. Methane also contributes to Titan's thermal inversion, and is primarily responsible for the stratospheres of the Giant Planets, less marked than on Titan due to the lower importance of aerosols.

3.2. Methane Photochemistry

The second implication of the discovery of new hydrocarbons in the Outer planets was the demonstration that their stratospheres are the place of active methane photochemistry. This process, also predicted prior to being observed Strobel (e.g. 1969), begins with the photolysis of methane ($CH_4 + h\nu \longrightarrow CH_3 + H$, $^1CH_2 + H + H$, $^3CH_2 + H + H$) which initiates a suite of chemical reactions leading to the production of numerous hydrocarbons. Fig. 5 shows an up-to-date view of Saturn's photochemistry (Moses et al. 2000) that was progressively constrained over the years by the detection of more hydrocarbons (so far CH_4, CH_3, C_2H_2, C_2H_4, C_2H_6, C_3H_8, CH_3C_2H, C_4H_2 and C_6H_6 have been detected). Similarly, although Uranus and Neptune were two faint and cold for their thermal emission to be measured in the 1970s, continuous technological developments and the use of space-borne facilities (particularly ISO and Spitzer) have by now provided outstanding spectra of these objects, revealing (to first order) a similar hydrocarbon composition as on Jupiter and Saturn (Fig. 6).

3.3. Equilibrium vs Disequilibrium Species

In addition to hydrocarbon emission, the Saturn spectrum of Fig. 4 shows absorption by ammonia (NH_3) and phosphine (PH_3). The detection of phosphine in Jupiter and Saturn (Gillett & Forest 1974; Bregman et al. 1975) marked the discovery of a new class of molecules. Although this was not realized immediately, phosphine is very different from ammonia. NH_3 is the dominant equilibrium form of nitrogen at all levels in Jupiter/Saturn tropospheres, allowing one, in principle, to determine the elemental

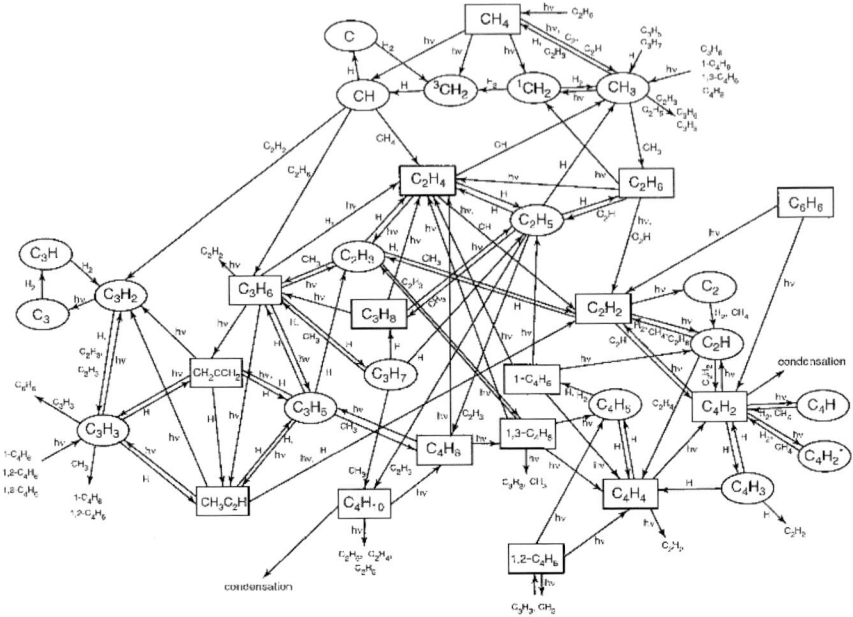

Figure 5. Main reactions in Saturn's methane photochemistry. Nine of the molecules have been already observed (see text). From Moses et al. (2000).

N/H ratio from the NH_3 mixing ratio. In contrast, under thermodynamical equilibrium, PH_3 should have a negligible abundance at observable levels (Barshay & Lewis 1978), and the dominant P-bearing species in the upper troposphere should be P_4O_6, itself ultimately converted to solid $NH_4H_2PO_4$ (Fig. 7). Thus, the large ($\sim 10^{-6}$) observed mixing ratio of PH_3 implies that the tropospheres of Jupiter and Saturn are out of equilibrium, and that phosphine is transported from deeper levels (where it is chemically stable) to upper levels (where it can be observed) more quickly than it is chemically destroyed. This concept of "disequilibrium species" was later refined to introduce the concept of "quench level" (e.g. Fegley & Prinn 1985), i.e. the level at which the characteristic time for vertical transport equals the chemical destruction rate of the considered molecule. Above the quench level, chemical reaction rates (whose kinetics decrease exponentially with decreasing temperature) are effectively frozen, and the atmosphere preserves the same abundance of the molecule as at the quench level. In the case of PH_3, the quench level occurs at ~ 1200 K; at this level, PH_3 is the dominant P-bearing species, meaning that the PH_3 abundance in the observable atmosphere does constrain the P/H ratio.

3.4. Physical Processes from Vertical Profiles

Measuring elemental ratios is one of the major goals of studying the Giant Planets, as they give constraints on the formation scenario. The above reasoning indicates that both equilibrium and disequilibirum species can potentially provide this information. This requires however that the associated molecules be observed at sufficiently deep levels so that other processes do not limit their apparent abundances. In this respect, the discovery of the 5-μm window of the Giant Planets was important, as this spectral region

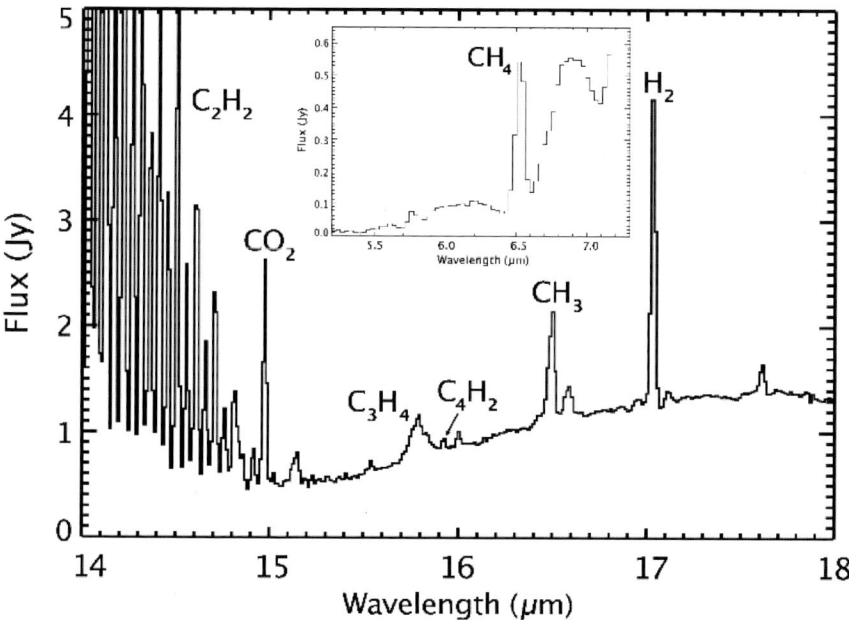

Figure 6. IRS/Spitzer observations of Neptune at 14-18 μm, showing the first detection of CH_3CCH and C_4H_2. The resolving power is 600. From Meadows et al. (2008).

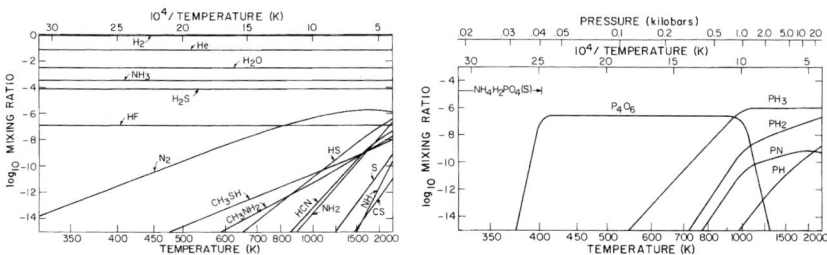

Figure 7. Equilibrium diagram for N- and S- species (left) and P-species (right) in Saturn's atmosphere. From Fegley & Prinn (1985).

(difficult to observe from the ground, and thus requiring high-altitude or airborne observatories – e.g. Larson et al. (1975)) shows little opacity by the major gases (H_2/He and CH_4) of Jupiter and Saturn and therefore sounds relatively deep levels (2-5 bar, depending on cloud cover) in their atmospheres. In these respects, highlight observations of Jupiter in the 1970's were (i) the measurement of ammonia and phosphine at 5 μm (ii) the discovery of (internal) water, with a highly sub-solar abundance ($\sim$ 1/50) at the probed levels (Larson et al. 1975) (iii) the detection of CO (Beer 1975), likely to be a disequilibrium species, and as expected much less abundant (by a factor 10^6) than methane.

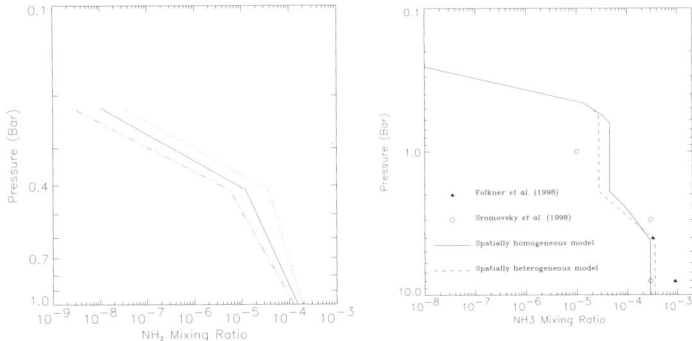

Figure 8. Vertical profile of ammonia in Jupiter based on 10-μm spectroscopy (left, sampling mean humidity conditions on Jupiter) and on 5-μm spectroscopy (right, representing "hot spots", i.e. particularly dry regions) from ISO. The NH_3 mixing at p >4 bar represents the deep abundance, the depletion over 1-0.4 bar is due to condensation, and the sharper depletion above 0.4 bar is due to photolysis. Note also how "hot spots" are depleted in NH_3 in the 1 bar region compared to mean conditions. From Fouchet et al. (2000).

Different wavelength ranges sound different atmospheric levels. The measurement e.g. of ammonia in Jupiter in several spectral domains (UV, 10-μm, 5-μm) allowed one to establish its vertical profile (see Fig. 8 for a more recent study making use of ISO), tracing several physical processes. In particular the NH_3 abundance shows two regions of depletion, one above ~1 bar due to condensation, and a second one above the tropopause (~400 mbar) due to photolysis. Below the ~4 bar level (probed at 5-μm), the NH_3 mixing ratio (~3.5×10^{-4}) presumably represents the bulk abundance, and indicates that the N/H ratio in Jupiter is somewhat (approximately 2.5 times) oversolar, a value that was essentially confirmed in 1995 by in situ measurements from Galileo (see below). These measurements also established that NH_3 shows large horizontal variability in its region of condensation, being strongly depleted in "hot spots", i.e. regions of low cloud cover and low humidity. Condensation and local meteorology also explains why the apparent abundance of oxygen is so low in Jupiter, as water vapor condenses at levels (5-10 bars in mean conditions, > 20 bars in hot spots) deeper than can be spectroscopically probed.

4. The Golden Age of Infrared

4.1. First Global Views of the Thermal Infrared Spectra

While ground-based observations in the telluric windows had demonstrated the promise of the thermal infrared range, spacecrafts launched in the 1970s and early 1980s provided the first complete views of these spectra, for all planets with substantial atmospheres. The most informative of these experiments were Fourier Transform Spectrometers with moderate (2-4 cm^{-1}) spectral resolution, such as IRIS/Mariner 9 (Mars, 1973), FTS/Venera 15 (Venus, 1975), and IRIS/Voyager 1 and 2 (Jupiter, 1979 and 1980; Saturn, 1980 and 1981; Titan 1980; Uranus, 1986; Neptune, 1989). Not only did these instruments open the spectral regions not accessible from the Earth (especially the 13-50 μm domain), but they also provided the first spatially resolved information.

At Mars and Venus, one of the major results was the retrieval of the atmospheric temperature field (and in the case of Mars, surface temperature) from the CO_2 15 μm band, illustrating the application to planetary sciences of inversion techniques developed for the Earth's atmosphere (e.g. Chahine 1970; Rodgers 1970) and revealing the nature of atmospheric circulation on these planets. Additional results involved the mapping of composition (H_2O and dust in Mars; H_2O, SO_2 and H_2SO_4 above Venus' clouds).

In terms of atmospheric composition, the most important outcome of the Voyager 1 exploration of the Giant Planets was probably the long-awaited measurement of helium. This was obtained by combining the emission spectrum in the far-IR (15-40 μm) range measured by IRIS – where the only source of opacity is the continuum-induced opacity of H_2 and He – with radio-occultation measurements. The principle is that the radio-occultation provides a measure of T/m at each level, where T is the atmospheric temperature and m is the mean molecular mass (see e.g. Conrath et al. (1991) for Neptune, with references for other planets). For various values of m (i.e. of the He/H_2 ratio) and hence various thermal profiles, synthetic spectra are calculated until the 15-40 μm brightness temperatures are matched. Complications may arise when clouds are present and affect the far-IR radiances, or if the radio-occultation is unreliable, and an alternative method, making use of the *shape* of the far-IR spectrum (that shows subtle variations with the He/H_2 ratio) has been devised (see Conrath & Gautier 2000). Results indicate that the He abundance (expressed as a He/H mass fraction) in Uranus and Neptune is within error bars consistent with the protosolar value, but the one in Jupiter and Saturn is *lower* (Fig. 9). For Jupiter, this result was confirmed by extremely accurate ($\pm 2 \%$) *in situ* measurements by the Galileo probe in 1995 (von Zahn et al. 1998). As hydrogen and helium have certainly been well mixed in the protosolar nebular (with no possible contribution of icy grains), the accepted interpretation is that helium undergoes a segregation at the ~Mbar level where metallic hydrogen form, and migrates towards the interior. This process, which leads to a depletion of the observable atmospheres of Jupiter and Saturn in helium, does not occur at Uranus and Neptune, where no metallic hydrogen transition is expected to occur.

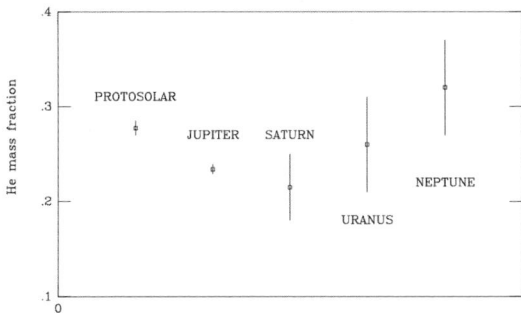

Figure 9. He/H mass ratio in the Giant Planets, compared to protosolar nebula value.

The other extremely important product of the measurement of the thermal IR spectrum of the Giant Planets is that it made possible to determine their total radiance (i.e. their effective temperature) and establish their energy balance. This revealed that with the remarkable exception of Uranus, Giant Planets emit more energy than they receive from the Sun (e.g. by as much as a factor 2.6 for Neptune), a consequence of their internal source.

If the Voyager observations did not provide many new molecular detections in the Giant Planets, they did, in contrast, revolutionize our understanding of Titan's chemistry. UV spectra demonstrated that N_2 was the major atmospheric species, and IR spectra (Fig. 10) revealed the presence of nitriles (HCN, HC_3N, C_2N_2, along with CH_3CN later discovered in ground-based millimeter-wave observations) in addition to hydrocarbons. Titan's atmosphere was then understood as the place of coupled photochemistry of CH_4 and N_2, leading to abundant formation of organic molecules, with the haze as the ultimate product.

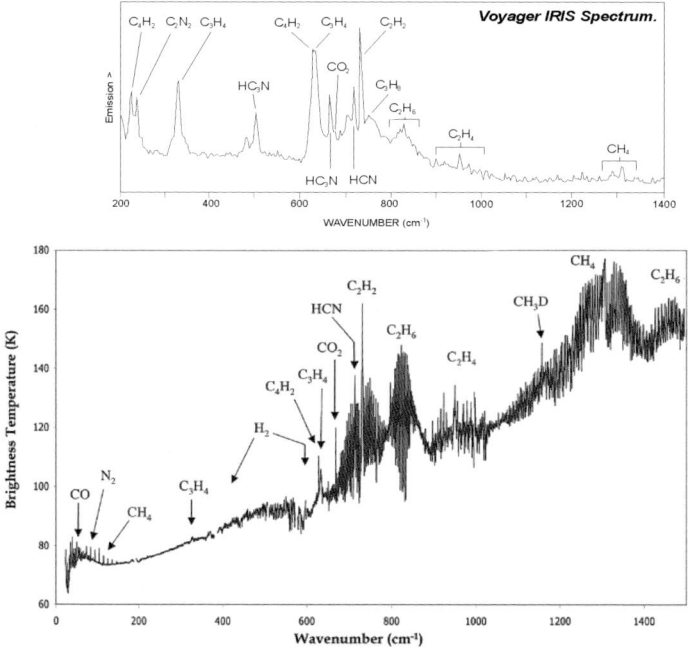

Figure 10. Titan: Voyager/IRIS spectrum (4.3 cm^{-1} resolution, limb viewing) and Cassini/CIRS spectrum (0.5 cm^{-1} resolution, nadir viewing).

4.2. Accumulating Molecules

The previous section illustrates that the 1970s and early 1980s were the key period for the first-shot exploration of planetary atmospheres, during which essential discoveries were made and physical concepts put in place. The IR had demonstrated its power, and from then on, the task remained to exploit it fully. The following 20 years or so witnessed the detection of many more molecules, providing a refined understanding of these atmospheres, and representing in this sense a golden age of the infrared. This was made possible by the combination of increasingly powerful instruments with increasingly large (4-10 m) telescopes. Instrumental performances improved in several directions in parallel: (i) spectral resolution (up to $R \sim 0.01$ cm^{-1}, i.e. a resolving power of

10^5 at 10 μm and 10^6 at 1 μm), thanks to the use of Fourier Transport Spectrometers or Fabry-Perots (ii) instantaneous spectral coverage, notably by using FTS (iii) sensitivity, achieved by cooling the instruments (or even the entire telescope, e.g. the space-borne ISO) to cryogenic temperature (iv) the possibility to combine spectrometers with an array of detectors, giving simultaneously access to the spatial dimension (1-D or 2-D) (iv) the development of novel techniques, such as heterodyne spectroscopy (at 10 μm and in the mm range), affording even higher spectral resolution. In what follows, we give examples of achievements and detections along with their physical implications.

4.2.1. Exploiting the 5-μm Window of Jupiter and Saturn

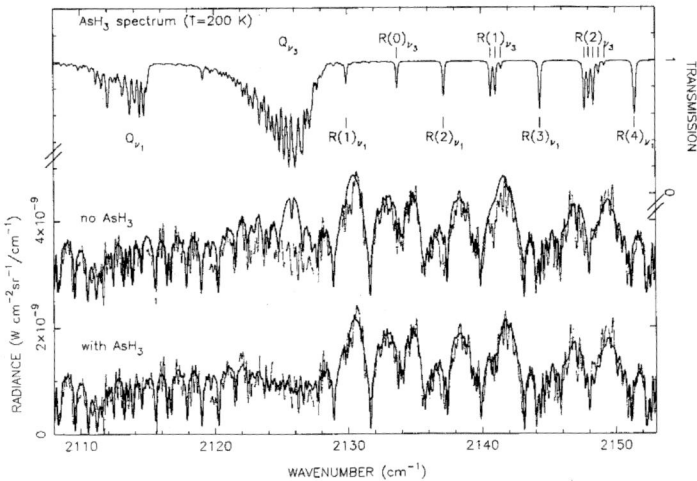

Figure 11. The detection of arsine in Saturn, obtained from CFHT/FTS spectra at a resolving power of 22000. From Bézard et al. (1990a).

The 1980 decade saw many observations of the 5-μm window of Jupiter and Saturn, permitting the discoveries of more disequilibrium species: CO on Saturn, GeH_4 (germane) and AsH_3 (arsine) in Jupiter and Saturn (see (Noll & Larson 1991; Bézard et al. 1990a) and references therein, and an example in Fig. 11). Similar to phosphine, the abundance of AsH_3 at 2-5 bar provides a measurement of As/H in Saturn's interior, found to be ~5 times solar. The case of germane was found to be somewhat different. The 2-5 bar abundance of GeH_4 is $\sim 4 \times 10^{-10}$, corresponding to a Ge/H ratio ~20 times less than solar. This, however, does not indicate that Saturn is globally depleted in germanium because thermochemical models indicate that at the quench level for GeH_4, the bulk of Ge is still contained in solid GeS and GeSe. CO is a complex case, having both an internal and external origin, but at least for Jupiter its measured abundance indicates that O is enriched with respect to solar (Bézard et al. 2002). Overall, these 5-μm observations indicated that Jupiter and Saturn are enriched in heavy elements (C, N, P, As and probably O) by factors of 2-10, with Saturn more enriched than Jupiter. This could be understood from the fact that the two planets have cores of similar masses, hence a similar budget of heavy elements, but very different masses of hydrogen (a factor ~3 larger at Jupiter).

Another major result of this epoch in the field of Giant Planets was the discovery of H_3^+ in Jupiter (Drossart et al. 1989), representing the first detection of this ion *in space*. This detection, which was followed by the observation of H_3^+ in Saturn and Uranus, was a major milestone in understanding the chemistry and dynamics of Giant Planet ionospheres (see chapters by Maillard and by Miller, this volume).

4.2.2. Deuterium in the Solar System

The deuterium abundance is a well known indicator of planetary formation and evolution. Measurements of the D/H ratio were obtained in the 1980s in the telluric (from HDO / H_2O) and outer (from CH_3D/CH_4, and later from HD/H_2) planets. In the outer solar system, evidence was found for two classes of objects (Owen et al. 1986): (i) Jupiter and Saturn, which have a D/H ratio representative of the protosolar value (ii) Uranus, Neptune, and Titan, which exhibit enhanced D/H values. Indeed, Uranus and Neptune may have been enriched in deuterium, during their formation, by the mixing of their atmospheres with comparatively larger cores containing D-rich icy grains. Similarly (but in a somewhat more complex manner), the D/H in Titan must reflect the combination of an initially high D/H in the primordial icy material that formed the satellite with the effect of fractionation processes in the atmosphere, such as methane photolysis and escape. Further support for this "two-reservoir" scenario was obtained when the D/H ratio was measured in comets and found to be even higher than in Uranus, Neptune, or Titan (Fig. 12). In the telluric planets, measurements indicate a large D/H enrichment in water compared to the terrestrial value of 1.5×10^{-4} (SMOW), by a factor ~ 5.5 in Mars and ~ 120 in Venus. This enrichment reflects the photolysis of water and preferential escape of deuterium, and suggests that Venus and Mars had once a much larger water endowment, but quantitative conclusions are difficult to draw.

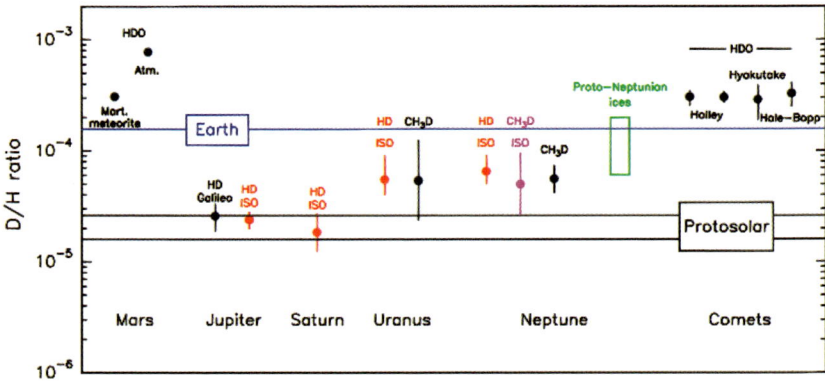

Figure 12. The D/H ratio in Mars, the four Giant Planets, and comets, compared to the protosolar value and to the telluric (SMOW) value.

4.2.3. Probing Below Venus' Clouds

In 1984, IR observations at the AAT (Allen & Crawford 1984) revealed that the nightside of Venus emitted measureable amounts of radiation in specific near-IR windows at 2.3 and 1.74 μm. These windows correspond to regions of very low CO_2 and H_2O opacity, allowing hot thermal radiation from the deep atmosphere, only partly filtered by cloud absorption, to leak out to space. Additional windows at 1.31, 1.27, 1.18, 1.0, 0.90 and 0.85 μm were subsequently discovered. Unlike dayside observations, which sample solar radiation reflected off Venus' clouds, these windows provide an unique opportunity to remotely probe the deep layers and even the surface of Venus. The 2.3 and 1.74 μm windows contain spectral signatures of several gases in the 15-45 km ranges, such as H_2O, HDO, SO_2, OCS, HF and HCl (Fig. 13). Although these gases had been detected before from in situ exploration by the Venera 12, Pioneer Venus and Vega descent probes, these spectroscopic observations (e.g. Bézard et al. 1990b; Pollack et al. 1993) allowed to solve conflicts between the reported abundances and vertical profiles of these species. In addition, they opened the way to the mapping of these species, first briefly during the Galileo flyby in 1990, and much more extensively by the Venus Express mission in 2006-2009, revealing e.g. latitudinal variations of CO related to an Hadley-type circulation.

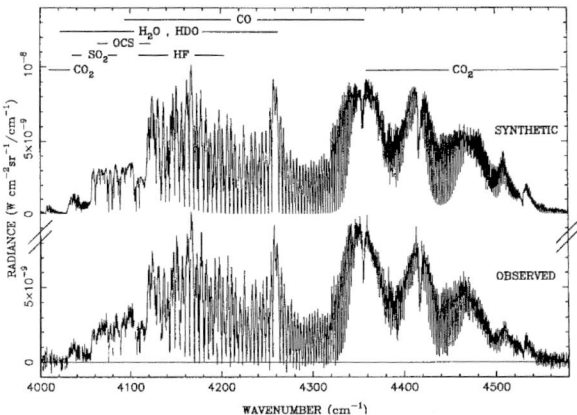

Figure 13. High-resolution spectrum (R=0.15 cm^{-1}) of Venus' 2.3 μm emission obtained from CFHT/FTS. Adapted from Bézard et al. (1990b).

4.2.4. External Sources of Material in Outer Planets

The operation of ISO in 1995-1998 provided many new discoveries on the composition of Outer Planet atmospheres (see review of Lellouch (1999)), particularly in the 10-50 μm part of the spectrum (and thanks notably to the then unsurpassed sensitivity of SWS). Perhaps the most significant was the detection of water in the *stratospheres* of all four Giant Planets and Titan (Fig. 14). As this water cannot originate from the interior due to the tropopause cold trap, it indicates an *external* source of oxygen. As originally discussed by (Feuchtgruber et al. 1997), the origin of this source could include (i) a permanent flux of interplanetary icy dust particles (ii) local sources originating from planetary environments (satellites, rings) (iii) discrete, cometary-type events delivering massive amounts of oxygen in the form of CO and H_2O. Evidence that comets are sources of material for atmospheres was proven in a spectacular way in 1994, when the impact of comet D/Shoemaker-Levy 9 with Jupiter induced major changes in the composition of Jupiter's stratosphere near the 0.1 mbar level. Newly detected and enhanced molecular species included CO, H_2O, S_2, CS_2, CS, OCS, H_2S, NH_3, HCN and C_2H_4 (see review of Lellouch (1996)). Although a number of these species were observed only for a short time (e.g. Fig. 14b), some have a long chemical lifetime ($\gg 1$ year, e.g. CO, CS, HCN and to a lesser extent H_2O) and are still observed to be present in Jupiter's atmosphere, with their spatial and vertical distribution evolving through diffusion and/or eddy mixing.

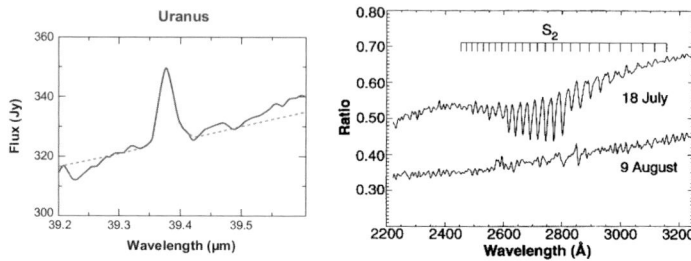

Figure 14. Evidence for external source of material in the outer planets. Left: detection of external water on Uranus from ISO/SWS (R = 1500). Adapted from Feuchtgruber et al. 1997. Right: detection (HST/FOS), and subsequent disparition of S_2 in Jupiter's atmosphere after the impact of comet Shoemaker-Levy 9. From Noll et al. (1995).

5. Concluding Remarks: Spacecraft Exploration

Knowledge of extrasolar planet atmospheres is nowadays probably as advanced as was the understanding of solar system atmospheres in the early seventies. Only a few molecules have been detected (H_2O, CH_4, etc.), but the physical principles are in place, and thanks to space telescopes (HST, Spitzer), the access to the entire spectrum is already possible. With new facilities becoming available in a few years (e.g. JWST, ELTs), one may expect that the knowledge of extrasolar atmospheres will soon, at least partly, catch up that of solar system bodies. However, *spacecraft exploration* will remain a specific feature of solar system science, and as a matter of conclusion, it may useful to briefly assess its uniqueness.

Spacecraft missions to the planets continue to carry out spectrometers, particularly in the IR and in the UV. However, given (i) the already advanced state of our knowledge of the planetary molecular inventories (ii) the fact that, due in part to mass and thermal constraints, ground-based spectrometers are usually more powerful (higher spectral resolution, better sensitivity) than their spaceborne counterparts (iii) the timescale (5-20 years) of implementation for spacecraft missions, making spectrometers in-flight "old technology" when they start producing data, it is perhaps not surprising that such spaceborne instruments now only occasionnally permit detections of new molecules, and that "first spectroscopic detections" are still obtained remotely. As an example, at the time of writing, and after 4 1/2 years in Saturn orbit, and in spite of a factor-of-10 gain in spectral resolution compared to its predecessor IRIS on Voyager (see Fig. 10), the CIRS spectrometer onboard Cassini has not provided any new detection in Titan's atmosphere (except for numerous isotopic variants of known molecules). Although exceptions do occur (e.g. the first detection of the OH radical by VIRTIS on Venus Express Piccioni et al. (2008)). the main interest of space-borne spectrometers is that, thanks to their uncomparable spatial resolution and limb sounding capabilities, they now provide us with three-dimensional views (vertical, latitudinal, and longitudinal) of atmospheric composition. By permitting a study, in this manner, the couplings beween composition and dynamics (see e.g. Teanby et al. (2008), for the case of Titan), they open up new fields of comparative planetology.

The uniqueness of spacecraft exploration also lies in the possibility of *in situ* sensing. This provides (i) access to complex molecules present in too low amounts to be detected spectroscopically (ii) access to species having no spectroscopic signatures (e.g. noble gases) (iii) unprecedented vertical resolution, particularly important for probing meteorological phenomena through the quickly-variable vertical profiles of condensing species. For such studies, the key instruments are certainly mass spectrometers. Recent examples of achievements can be found with the operation of the Galileo (Jupiter) and Huygens (Titan) probes. At Jupiter, Galileo has revealed that the level of enrichment over solar abundances is strinkingly similar (a factor of ~3) for most measured elements (C, N, S, Ar, Kr, Xe; O cannot yet be measured) (Owen et al. 1999). This is a strong constraint for formation models, though there is still debate on the interpretation. At Titan, in addition to isotopic and elemental measurements, the Huygens GCMS provided a vertical profile of methane with exquisite accuracy and vertical resolution (Niemann et al. 2005), permitting, in combination with local temperature/pressure measurements, fine meteorological analyses and evidence for methane drizzle (Tokano et al. 2006). No less spectacular are the current results from the Cassini/INMS (ion-neutral mass spectrometer), which, by sampling Titan's upper atmosphere near 1000 km at each of the Titan flybys of Cassini, have revolutionized our understanding of Titan's atmosphere chemistry and revealed a degree of molecular and ionic complexity far exceeding expectations (Vuitton et al. 2007). This kind of achievements, for any predictable future, won't be possible in extrasolar planetary science.

Acknowledgments. I thank Dale Cruikshank for discussion of Kuiper's and Moroz's achievements.

References

Adams, W.S., & Dunham, T., Jr. 1932, PASP, 44, 243
Allen, D.A., & Crawford, J.W. 1984, Nat, 307, 222
Barshay, S.S, & Lewis, J.S., 1978, Icarus, 33, 593
Barth, C.A., & Hord, C.W. 1971, Sci, 173, 3393
Beer, R., 1975, ApJ, 200, 167
Bézard, B., et al. 1990, ApJ, 346, 509
Bézard, B., de Bergh, C., Crisp, D., & Maillard, J.P., 1990, Nat, 345, 508
Bézard, B., et al. 2002, Icarus, 159, 95
Bregman, J.D., Lester, D.F., & Rank, D.M., 1975, ApJ, 202, L55
Carleton, N.P., & Traub, W.A., 1972, Sci, 177, 988
Chahine, M.T., 1970, J. Atmos. Sci., 27, 960
Clancy, R.T. et al., 2004, Icarus, 168, 116
Connes, P, Connes, J., Benedict, C., & Kaplan, L.D., 1967, ApJ, 147, 1230
Conrath, B.J. et al. 1991, JGR, 96, 18907
Conrath, B.J., & Gautier, D., 2000, Icarus, 144, 124
Danielson, R.E., Caldwell, J.J., & Larach, D.R., 1973, Icarus, 20, 437
Drossart, P. et al., 1989, Nat, 340, 539
Encrenaz, T. et al. 2004, Icarus, 170, 424
Fegley, B., Jr. & Prinn, R.G., 1985, ApJ, 299, 1067
Feuchtgruber, H., et al. 1997, Nat, 389, 159
Fouchet, T., et al., 2000, Icarus, 143, 223

Gillett, F.C., 1975, ApJ, 201, L41
Gillett, F.C., & Forest, W.J., 1974, ApJ, 187, L37
Herzberg, G., 1952, ApJ, 115, 337
Kaplan, L.D., Connes, J., & Connes, P., 1969, ApJ, 157, L187
Kuiper, G.P., 1944, ApJ, 100, 378
Kuiper, G.P., 1947, ApJ, 106, 251
Kuiper, G.P., 1949, ApJ, 109, 540
Larson, H.P., Fink, U., Treffers, R. & Gautier, T.N. III, 1975, ApJ, 197, L137
Lellouch, E., 1996. In "The Collision of Comet Shoemaker-Levy 9 and Jupiter, IAU Colloquium 156", (Cambridge University Press), p. 213.
Lellouch, E., 1999. In "The Universe as Seen by ISO". ESA-SP 427., 125
Lindal, G.F., et al. 1983, Icarus, 53, 348
Meadows, V.S. et al. 2008, Icarus, 197, 585
Moroz, V.I., 1964, Sov. Astron., 8, 273
Moses, J.I., et al. 2000, Icarus, 143, 244
Niemann, H.B., et al. 2005, Nat, 438, 779
Noll, K.S., & Larson, H.P., 1991, Icarus, 89, 168
Noll, K.S., et al., 1995, Sci, 267, 1307
Owen, T., Lutz, B.L., & de Bergh, C., 1986, Nat, 320, 244
Owen, T., et al., 1999, Nat, 402, 269
Piccioni, G. et al., 2008, A & A, 423, L29
Pollack, J.B., et al., 1993, Icarus, 103, 1
Rea, D.G., O'Leary, B.T., & Sinton, W.M., 1965, Sci, 147, 1286
Rodgers, C.D., 1970, Quat. J. Roy. Met. Soc., 96, 654
Sinton, W.M., 1957, ApJ, 126, 2318
Sinton, W.M., & Strong, J., 1960, ApJ, 131, 459
Smith, M.D, 2002, JGR, 107, DOI 10.1029/2001JE001522
Spinrad, H., 1963a, ApJ, 138, 1242
Spinrad, H., 1963b, ApJ, 137, 1319
Strobel, D.F., 1969, J. Atm. Sci, 26, 906
Teanby, N.A. et al., 2008, JGR, 113, 10.1029/2008JE003218
Tokano, T. et al., 2006, Nat, 442, 432
Trafton, L.M., 1975, Icarus, 24, 443
Vuitton, V., Yelle, R.V., & McEwan, M.J., 2007, Icarus, 191, 722
Wildt, R., 1932, Ve. Goe., 2, 216
von Zahn, U., Hunten, D.M., & Lehmacher, G., 1998, JGR, 103, 22815

The Molecular Ion H_3^+ in Emission in Planetary Atmospheres

J.P. Maillard

Institut d'Astrophysique de Paris, 98b Blvd Arago, 75014 Paris, France

S. Miller

University College London, Gower Street, London, WC1E 6BT, United Kingdom

Abstract. The molecular ion H_3^+ was detected by spectroscopy twenty years ago, for the first time outside of the laboratory, in the upper atmosphere of the polar zones of Jupiter. This detection made possible temperature, abundance measurements, imaging of the ionic emission, and triggered its search in the atmosphere of the other giant planets, leading to a positive detection in Saturn and Uranus. These measurements, mainly in the ν_2 band around 3.7 μm and its overtone $2\nu_2$ at 2.1 μm, revealed a wealth of information on the planetary magnetospheres and the auroral phenomenon. On the hot Jupiters, H_3^+, likely excited through EUV radiation, could be an important target to prove the existence of a gaseous planet and to monitor the escape processes of the atmosphere. The attempts of detection in exoplanets have been so far unsuccessful with the current limits of detection. From the experience gained on the role played by this ion in the energy balance of the giant planets of the Solar System, can be inferred its role of thermostat in the upper atmosphere and the ionosphere of giant extrasolar planets as main cooling agent contributing to their stability. However, the distance to the star is an important parameter for H_3^+ to be able to form. Occultation spectroscopy with more transiting planets known should be the most promising method for this search.

1. Introduction

We live in a universe dominated by hydrogen. The atmospheres of hot stars, whatever their metallicity, are dominated by atomic hydrogen, along with protons and, especially for low metallicity stars, the H^- ion. The interstellar medium and the clouds of which it is composed are dominated by the H_2 molecule. Next in the series would be the triatomic hydrogen molecule, H_3. But this species is unstable, and it falls to the triatomic hydrogen ion, H_3^+, to form the most fundamental, stable, polyatomic molecule. H_3^+ was discovered by Thomson (1911) in the course of his experiments on "positive rays". During the next two decades there were numerous attempts to demonstrate the stability of triatomic hydrogen molecules in various stages of ionisation until was demonstrated that a triangular, three-centre bonded form of H_3^+ would be stable. Hirschfelder (1938) pointed out that it would have infrared-active vibrational modes, although his geometry, intermediate between a right-angled and an equilateral triangle, meant that he predicted two infrared-active vibrations.

From then on, attempts were made to measure the H_3^+ spectrum in the laboratory (Oka 1980) and elsewhere. The initial focus of searches for H_3^+ outside of the laboratory was on the dense clouds subject to cosmic ray ionisation, following suggestions in the 1960s (Martin et al. 1961), and the incorporation of H_3^+ into schema for gas-phase chemistry (Herbst & Klemperer 1973) that aimed to explain the abundance of complex interstellar molecules discovered by radio astronomy (Rank et al. 1971). However, it was not until 1996 that H_3^+ was first discovered in interstellar clouds (Geballe & Oka 1996).

The first discovery in space came from the unexpected detection twenty years ago (Drossart et al. 1989) of H_3^+ in the infrared spectrum of Jupiter's aurorae. This observation was the origin of more observations of the giant planet, by high resolution spectroscopy and by spectral imaging, to study the production processes of the ion, the morphology of its distribution, its abundance, its intensity variability and its link to other indicators of auroral activity in the UV and in the mid-infrared. Through all these studies H_3^+ appeared as a crucial probe to determine the properties of the jovian magnetic field and its interaction with the main Jupiter's moons, also as an important contributor to the thermal structure of the upper atmosphere, and a precious indicator to test the upper atmospheric models. In parallel, laboratory studies and theoretical modelling of the molecular ion have developed for a better knowledge of its spectrum in all the infrared range. Naturally, H_3^+ was also searched for on the other giant planets of the Solar System. On the other hand, soon after the report of the first giant planet outside of the Solar System (Mayor & Queloz 1995), the possibility to detect this ion in the atmosphere of exoplanets was proposed (Miller et al. 2000).

The current paper briefly presents, from an astronomical point of view, the production and destruction mechanism of H_3^+, its key spectroscopic properties, the main observations made in Jupiter, then in Saturn and Uranus, and in summary the main differences of H_3^+ emission origin in the giant planets. Next, the current attempts of detection in extrasolar giant planets, still unsuccessful, are reviewed. The second part of the paper is devoted to the role of H_3^+ in the planetary atmospheres, including these new targets, in particular in their stability, to conclude on the most appropriate strategy to expect a positive detection.

2. Production and Destruction of H_3^+

The molecular ion H_3^+ is produced wherever molecular hydrogen coincides with ionising radiation or particles. Following the ionisation of H_2 to H_2^+, H_3^+ is produced through the rapid ion-neutral reaction:

$$H_2^+ + H_2 \rightarrow H_3^+ + H. \tag{1}$$

H_2^+ can be formed by several chemical reactions, which depend on the particular physical environment. In a plasma, or in the interstellar medium (ISM) with the bombardment of cosmic rays, or from energetic electrons precipitating along magnetic field lines, H_2^+ is produced by:

$$H_2 + e^- \rightarrow H_2^+ + 2e^-. \tag{2}$$

Another possibility comes from the presence of an extreme ultraviolet (EUV) radiation giving the reaction:

$$H_2 + h\nu \rightarrow H_2^+ + e^-. \tag{3}$$

Another important reaction is thought to be that of charge exchange:

$$H_2 \, (v \geq 4) + H^+ \rightarrow H_2^+ + H, \tag{4}$$

although the rate constant for this reaction is still subject to some considerable doubt. Note that the difference in ionisation energy of H and H_2 is provided by the reaction requiring vibrationally excited H_2 molecules, with $v \geq 4$.

The production of H_3^+ is balanced by the destruction reaction:

$$H_3^+ + X \rightarrow HX^+ + H_2, \qquad (5)$$

where H_3^+ acts as an universal protonator. This type of reaction initiates a myriad of chain reactions which is the basis of interstellar medium chemistry with X = O, CO, HD, H_2O, CH_4 .... In the upper atmosphere of the giant planets the destruction of H_3^+ is made through several reactions with different branching ratios, the two first reactions being the most probable:

$$H_3^+ + e^- \rightarrow H + H + H \qquad (6)$$
$$\rightarrow H_2 + H \qquad (7)$$
$$\rightarrow H_3. \qquad (8)$$

All these reactions have been introduced in models to predict the production and destruction of H_3^+ in the ISM (Herbst & Klemperer 1973), in the giant planets of the Solar System (Achilleos et al. 1998) and in the extrasolar giant planets (Yelle 2004).

3. Laboratory Spectroscopy of H_3^+

H_3^+ is the simplest polyatomic molecule, consisting of only three protons and two electrons. Its ground state electronic configuration, X^1A_1', has a dissociation energy of ~ 4.3 electron-volts, and shows a remarkable equilateral triangle, D_{3h}, equilibrium geometry (Fig. 1) (Meyer et al. 1986). The first excited state, $A^3\Sigma_u^+$ involves promoting one of the two bonding electrons to an anti-bonding orbit, and this state is bound by only ~ 0.15 eV in its ground vibrational state (Alijah & Varandas 2006). This means that, effectively, H_3^+ has no electronic spectrum. Its D_{3h} symmetry also means that the molecule has no permanent dipole. Thus, it has no *allowed* microwave or submillimetre pure rotational spectrum; although *forbidden* rotational transition frequencies and line strengths have been calculated (Pan & Oka 1986), they have never been observed.

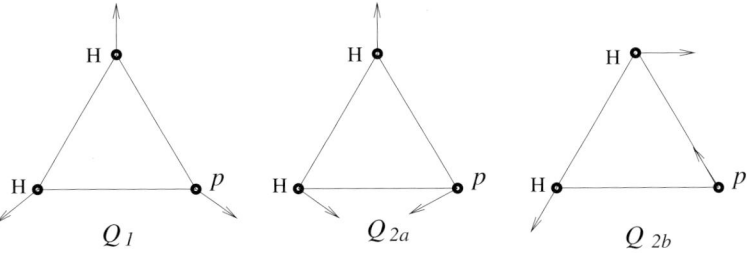

Figure 1.   Geometry of H_3^+ : equilibrium structure and normal modes.

The initial detection of the infrared H_3^+ emission spectrum in the ν_2 band was obtained from a discharge cell in a flow of pure H_2 (Oka 1980). Since this early work the

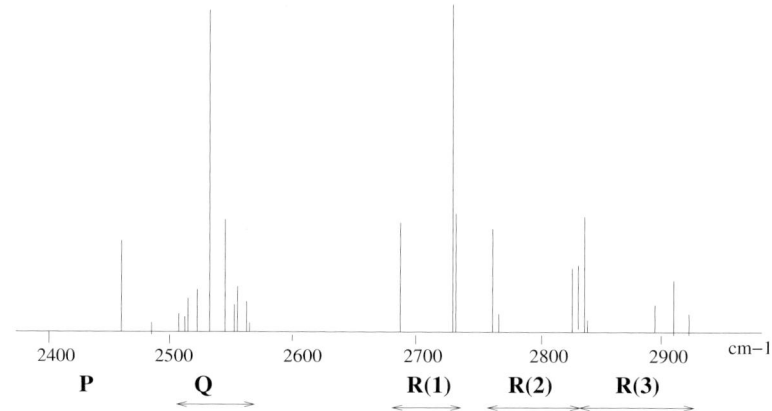

Figure 2. Theoretical spectrum of the ν_2 fundamental band of H_3^+ assuming a temperature of 200 K. The absence of an R(0) line is the proof that all three protons are equivalent.

laboratory study of H_3^+ has continued with different spectroscopic techniques, with discharge in flows of H_2 and helium, making possible to identify lines of the fundamental band, its overtones, a variety of hot and combination bands. An important step was the overtone spectrum published by Majewski et al. (1989).

The equilateral triangle shape of H_3^+ gives rise to two vibrational modes, the symmetric stretch (breathing) mode ν_1 at $3185\,\text{cm}^{-1}$ and ν_2 asymmetric stretch/bend at $2521\,\text{cm}^{-1}$. Of these, only the ν_2 mode is active. According to the spectroscopic rules associated with the normal harmonic oscillator, rigid rotor model of ro-vibrational transitions, the only infrared-allowed bands, involve changes of the ν_2 quantum number, ν_2, of 1. However, H_3^+ is an extremely anharmonic vibrator, which earns it the sobriquet "floppy" (Sutcliffe et al. 1987). It means that accurate calculation of the line frequencies and of the line intensities which are needed to derive temperatures and ion densities, requires an *ab initio*, "first-principles", fully-coupled quantum calculation if useful numbers are to be produced (Meyer et al. 1986; Miller & Tennyson 1987; Miller et al. 1989).

The $2\nu_2$ ($\nu_2 = 2 \rightarrow 0$) overtone (Drossart et al. 1989), the $3\nu_2$ ($\nu_2 = 3 \rightarrow 0$) second overtone (Lee et al. 1991) and several hot bands (Bawendi et al. 1990; Raynaud et al. 2004) are all known in the laboratory and in space, along with the forbidden fundamental and hot ν_1 bands ($\nu_1 = 1 \rightarrow \nu_1 = 0$) and ($\nu_1 = 1, \nu_2 = 1 \rightarrow \nu_1 = 0, \nu_2 = 1$) (Xu et al. 1992). The $\nu_1 \rightarrow \nu_2$ difference band has also been calculated (Miller and Tennyson, unpublished) but not yet measured. Currently, about 900 transitions have been measured and assigned, from 1546 to 12,419 cm^{-1}, which are reported in a specific H_3^+ database *The H_3^+ resource center* (Lindsay & McCall 2001). A complete review of laboratory spectroscopy of H_3^+ can be found in McCall (2000). The ν_2 fundamental band is by far the strongest spectral signature of H_3^+ (Fig. 2), and the vibration-rotation bands $1 \rightarrow 0, 2 \rightarrow 0$ provide the most observable bands for astronomical purposes.

4. Detection of H_3^+ in the Giant Planets of the Solar System

4.1. H_3^+ in Jupiter's Polar Regions

The detection of H_3^+ in Jupiter came from the study of the auroral activity in the giant planet. Two spots of strong atomic and molecular hydrogen emissions had been observed in the polar zones of Jupiter by the Voyagers UV spectrometer (Broadfoot et al. 1979), then, confirmed by IUE observations. From their location these emissions were interpreted as signatures of an intense auroral activity. Later, to confirm the prediction made on the excitation of the quadrupole lines of H_2 on the auroral regions (Kim et al. 1985) jovian spectra between 2 and 2.2 μm were recorded with a 5 arcsec aperture projected at high latitude on the planetary disk, with the Fourier Transform Spectrometer (FTS) at the CFH Telescope (Maillard & Michel 1982). The observations took place in 1988, Sept. 24. They allowed a firm detection of H_2 ro-vibrational lines from the S1(0), S1(1), and S1(2) lines (Kim et al. 1990) with in addition, a set of more than twenty unidentified lines. From the comparison with a laboratory spectrum of a discharge through H_2, recorded at the Herzberg Institute of Astrophysics (Majewski et al. 1989) and the result of *ab initio* calculations (Miller et al. 1990) the unknown lines were definitely assigned as belonging to the $2\nu_2$ band of H_3^+ (Drossart et al. 1989).

Following the detection of H_3^+ in the southern auroral zone from its $2\nu_2$ band the observations were conducted with the same instrument to observe the fundamental band at 4 μm at high spectral resolution (Maillard et al. 1990). The run took place on 1989, Nov. 10 and 13 (UT). Up to 42 lines of the ν_2 band in emission were detected, in the range 2400 - 2900 cm^{-1}, at a resolution of 15,000 (Fig. 3). Remarkably, a pure H_3^+ spectrum was observed, what was not possible in laboratory, on top of a negligible continuum level. This aspect was explained by the production of H_3^+ in the upper jovian atmosphere, at the 1 - 100 nbar level, where the concentration of methane is low. While the H_3^+ 1 - 0 band is coincident with the strong ν_3 band of CH_4, the solar reflected flux and the internal emission are almost totally absorbed leaving just the H_3^+ lines.

A rotational temperature was derived for the southern and northern zones, respectively of 1000 ± 36 K and 835 ± 50 K. The intensity of the lines was on the average two times stronger in the south than in the north. The hot band $2\nu_2 - \nu_2$ was not detectable, compatible with the estimated rotational temperature. A thermal mechanism for the production of H_3^+ was implied. Later, a narrower portion of the same band was observed at much higher resolution (Fig. 4), still with the CFHT-FTS, for dynamics studies of the ionosphere (Maillard et al. 1999).

After the publication of the 1990 spectrum, it was realized that at 3.53 μm (or 2830 cm^{-1}) a multiplet of H_3^+ emission lines (Fig. 3) above a low and clean continuum level would be easy to isolate with a narrow-band filter, to image the auroral zones in the molecular ion. A series of images of Jupiter recorded in this band revealed the spatial distribution of H_3^+ emission in the jovian ionosphere as seen from Earth (Baron et al. 1991; Drossart et al. 1992). This type of study made possible to compare the H_3^+ distribution to the images of the auroral activity recorded in the UV in the H_2 Werner band (Fig. 5) and at longer infrared wavelength.

4.2. H_3^+ in Jupiter's equatorial zone and mid-to-low latitudes

H_3^+ has also been observed outside of the polar regions, in the north equatorial belt (NEB) and in the across the planetary disk. The observation in the NEB was made with the CFHT-FTS at a resolution of about 25,000, making possible the identification of

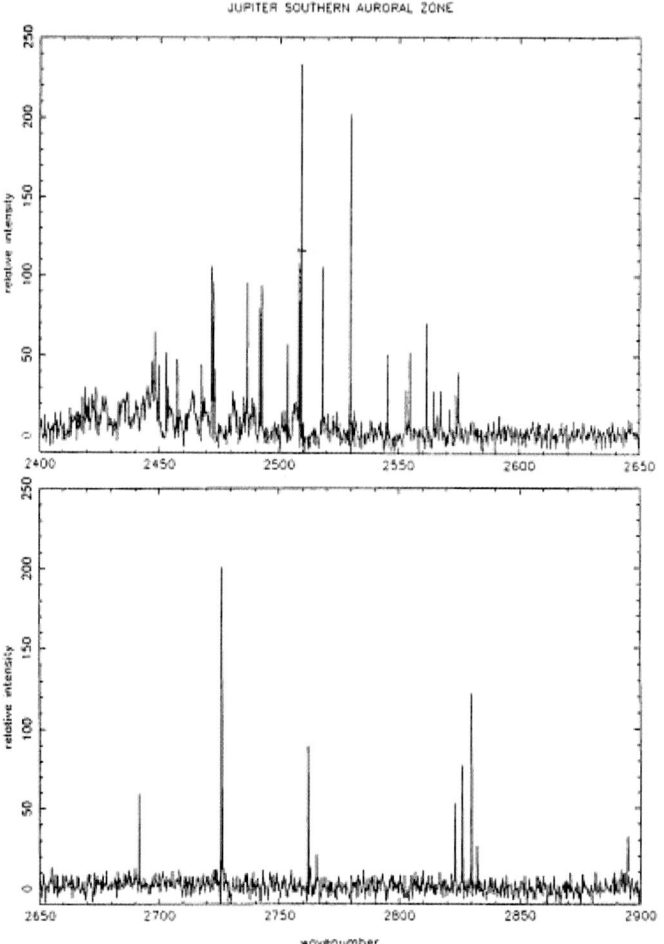

Figure 3. First high-resolution spectrum (R ≃ 15,000) of H_3^+ $1 \leftarrow 0$ in the southern auroral zone of Jupiter obtained with the CFHT-FTS (Maillard et al. 1990).

$^{13}CH_4$ lines and several weak H_3^+ lines (Marten et al. 1994). The NEB lines correspond to a column density about 10 times smaller than in the auroral regions and a temperature in average colder, at 800 K ± 100, implying a different mechanism of production than in the polar regions, with considerable temperature variations depending on the exact location on the planet.

The 60-arcsec long slit of the CGS4 facility spectrograph on the United Kingdom Infrared Telescope (UKIRT) was aligned with the jovian CML, in order to get high-spatial, high-spectral resolution latitudinal profiles of the jovian H_3^+ emission (Lam et al. 1997; Miller et al. 1997). The papers concluded that transport of H_3^+ ions from the auroral regions equatorward would not be sufficient to explain the densities of the ion observed at mid-to-low latitudes, and the spatial distribution of emission suggested

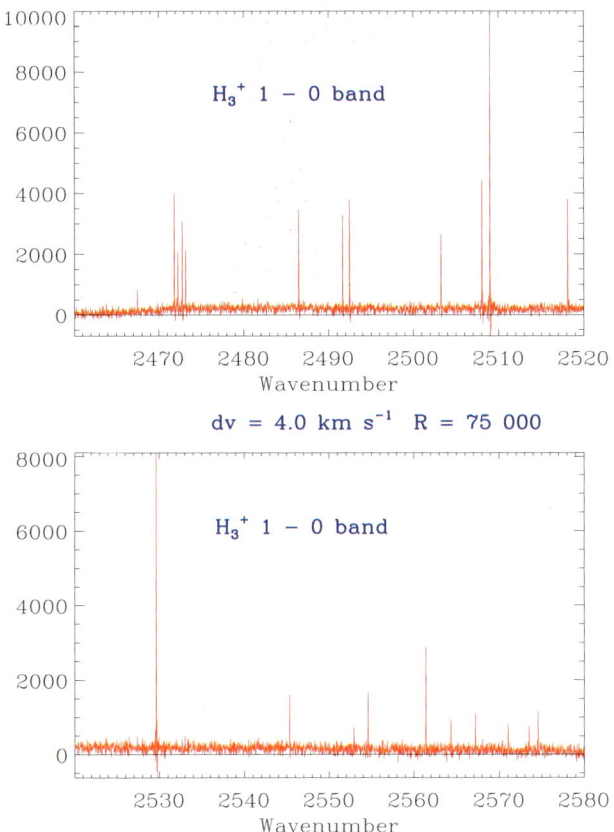

Figure 4. Small portion of the H_3^+ $1 \leftarrow 0$ band in the Q branch, at high resolution (R = 75,000), in the southern auroral zone of Jupiter (aperture 2.5") obtained with the CFHT-FTS in October, 1998.

that it was being produced *in situ* by a low-level flux of particle precipitation, probably westward drifting electrons of energies lower than those producing the brightest auroral emission. Follow up studies using the 30-arcsec long slit of the CSHELL facility spectrograph at the NASA Infrared Telescope (IRTF) (Rego et al. 2000) confirmed the need for ionisation sources other than solar EUV to explain the mid-to-low latitude emission. Magnetically, the mid-to-low latitudes connect to regions in the magnetosphere inside the orbit of Io. This magnetospheric region includes the main jovian radiation belts. But the nature of the ionisation sources that produce the mid-to-low latitude emission is still poorly understood, despite attempts to model particle precipitation from sources inside Io's orbit (Abel & Thorne 2003).

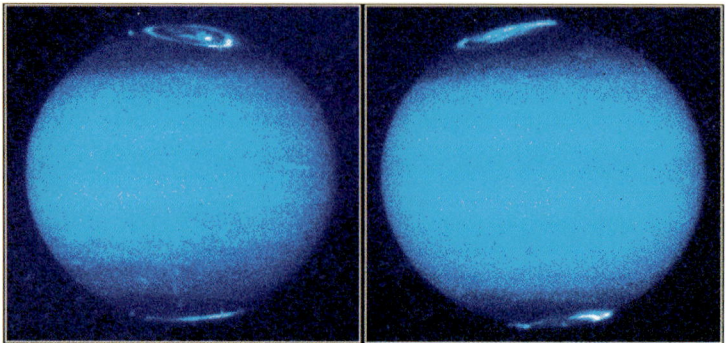

Figure 5. The auroral zones of Jupiter imaged by the WFPC2 camera on HST in the Werner bands of H_2 (UV) *(Credit: J. Clarke, U. of Mich.)*

4.3. H_3^+ in the Polar Regions of Saturn and Uranus

The detection of H_3^+ in the polar regions of Jupiter justified looking for the ion in the other giant planets. A strong H_3^+ emission was detected in Uranus with CGS4 (Trafton et al. 1993), a medium resolution spectrum showing 11 emission features of the ν_2 band between 3.89 and 4.09 μm, mainly from the Q-branch (Fig. 6).

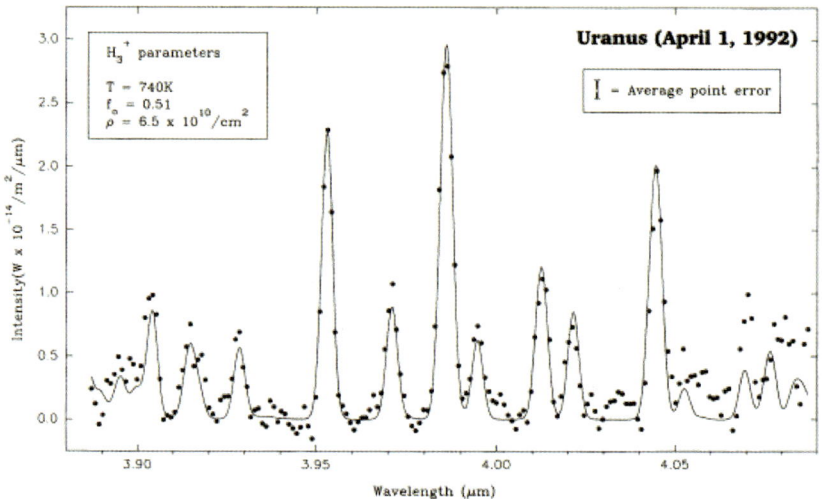

Figure 6. Spectrum of Uranus (filled circles) compared with the fit obtained using the model including 1 - 0 H_3^+ emission with the parameters given on the figure.

With the same instrument a weak H_3^+ emission of the ν_2 was detected at the poles of Saturn by three emission lines, two orders of magnitude lower than the column density at the south pole of Jupiter (Geballe et al. 1993). More recently, the southern Saturn's aurora was imaged from ground at 3.953 μm (Stallard et al. 2008a). The molecular ion is not detected in Neptune.

4.4. Difference of H_3^+ Emission origin

From all these observations of H_3^+, conclusions can be drawn on the mechanisms of emission. The emission is mainly concentrated in the auroral zone where the molecular ion is formed by charged particles streaming along the planet's magnetic field lines in to the upper atmosphere (between 1 and 100 nbar). This is the main mechanism to produce H_3^+ in the giant planets of the Solar System. Correlation with the solar cycle activity has been clearly demonstrated in Uranus (Encrenaz et al. 2003; Trafton et al. 1999). At Jupiter, the aurorae are formed through interactions with the solar wind and particles released by Io's volcanoes. The emission of H_3^+ is strong at Jupiter, moderately strong at Uranus and weak at Saturn, not following the relative strength of the magnetic fields (almost ten times weaker in Uranus than in Saturn). This major difference may be due to the different efficiency of H_3^+ destruction mechanism by the hydrocarbons in the atmosphere, which depends on the altitude of the homopause.

Production by the solar EUV flux, the other production mechanism, on the H_2-rich atmosphere (Eq. 3), plays a minor role and has only been observable in the weak emission on the Jupiter's mid-to-low latitude zones (Sect. 4.2).

5. Detection of H_3^+ in Giant Exoplanets

In the past decade, several hundreds of planets have been discovered orbiting stars in the neighbourhood of our Sun, of which a large portion are giant planets with a mass of about one to 10 Jupiter's masses. Of these, more than half have been found with semi-major rotation axes of less than 1 AU, and about one hundred orbit at 0.1 AU or less. Several reasons have been put forward to attempt to detect H_3^+ in the giant exoplanets. Already, soon after the detection of the first exoplanet around the solar-type star 51 Peg (Mayor & Queloz 1995) came the idea that H_3^+ could be the first species to detect, which without a doubt would be proving the presence of a gaseous giant planet. As a matter of proof, observations of HD209458b have indicated that the planet is surrounded by an expanded atmosphere of atomic hydrogen that is escaping hydrodynamically (Vidal-Madjar et al. 2003). It could be inferred that their lower atmosphere is mainly made of molecular hydrogen, which is exposed to a strong EUV flux. H_3^+ should be excited following the reactions given by Eq. 3 and Eq. 1. Moreover, since the H_3^+ spectrum is in emission, the contrast between it and the stellar flux can be very favorable, provided high resolution spectroscopy is used.

Possible abundance of H_3^+ in the upper atmosphere of extrasolar giant planets with semi-major axis from 0.01 to 0.1 AU has been estimated by a model (Yelle 2004). Hot Jupiters are predicted to have up to 10^5 times Jupiter's H_3^+ emission because they experience strong UV irradiation. The model shows also that even though the escape rates are large, the atmospheres are stable over time scales of billions of years.

5.1. Programs of Detection

Several programs aiming for the detection of H_3^+ towards stars with an exoplanet have been launched. So far, the results have been negative. A claimed positive detection of Q(1,0) and Q(3,0) was reported from a young Herbig AeBe star, HD14159, as formed in the protoplanetary disk around the star (Brittain & Rettig 2002). This detection was not confirmed from a deep search at the Subaru telescope with the IRCS spectrometer of the strongest lines of the ν_2 band in the 3.22 to 4.01 μm range (Goto et al. 2005).

The most complete program to date has been conducted to search for emission from the Q(1,0) transition of H_3^+ at 3.953 μm from extrasolar planets orbiting six late-type dwarfs, with CSHELL, which has a nominal resolving power, $\lambda/\Delta\lambda$, of ~ 40,000, equivalent to 7.5 km s^{-1}. A limit on the H_3^+ emission from each system was set. The most stringent limit of 6.3×10^{17} W was obtained for GJ 436, a close M2.5 star (10.2 pc) (Shkolnik et al. 2006). Finally, a new upper limit on the luminosity of H_3^+ emission from the exoplanet host τ Boo, target of the previous program, has been reported (Laughlin et al. 2008). Using high-resolution near-infrared spectra data collected with CSHELL, with a new limit of 9.0×10^{17} W which pushes down the previous limit by a factor of approximately 2.5.

6. The Role of H_3^+ in the Planetary Atmospheres

6.1. The H_3^+ "Thermostat" in Solar System Giant Planets

The temperature profiles of planetary atmospheres are controlled by a balance of a number of heating and cooling processes. In the upper atmosphere, above the homopause, the main heating processes are due to absorption of stellar EUV radiation and particle precipitation from the magnetosphere, both of which are *external* heating processes, and to *internal* heating such as Joule heating and ion drag. We will discuss the role of H_3^+ in the internal heating processes in the next sub-section. The main cooling mechanisms involve downward conduction and radiation to space, and horizontal transport of hot air by wind systems.

By definition, the homopause is the point at which convective mixing of the atmosphere is no longer important, and individual species have their own scale height, H_a:

$$H_a = kT/m_a g \qquad (9)$$

where the label a refers to an individual species, such as H, H_2, H_3^+, He, CH_4 ..., T is the temperature, m is the mass, and g the acceleration due to gravity. Since the local density of any atmospheric species falls off with altitude h approximately as:

$$N_a(h) = N_a(0)\exp[-h/H_a] \qquad (10)$$

it is clear that heavier species, such as the hydrocarbons, are confined to the very lowest levels of the upper atmosphere, and are concentrated below, or just above, the homopause.

Since H_3^+ reacts with, and is destroyed by hydrocarbons, it can therefore only play a role above the homopause. For Jupiter, that confines it to altitudes ~ 300 km above the 1-bar level, or higher, where the pressure is of the order of 1 μbar, or less, and the number density of H_2 is generally less than 10^{20} m^{-3}. This raises the issue as to whether H_3^+ levels will be populated according to local thermodynamic equilibrium (LTE), since the times between collisions become increasingly longer with increasing altitude/decreasing number density. Note that the radiative lifetime of H_3^+ vibrational levels is of the order of 0.01 s (Dinelli et al. 1992). Within a vibrational band, however, rotational sub-levels are generally in LTE. This makes it possible to derive accurate thermospheric temperatures, from the ratio between the intensities of ro-vibrational lines that have the same upper vibrational quantum numbers. But overall column densities derived from the absolute intensities of the lines have to take non-LTE effects into consideration (Melin et al. 2005).

At low gas densities, downward conduction is not very efficient as a cooling mechanism (Melin et al. 2006). Bulk transport of hot air by winds may be important in the transport of heat from regions of high energy input and generation such as the auroral/polar regions; again, this will be discussed in the next sub-section. That leaves in the first approximation, just radiation to space. The main components of the neutral thermosphere, above the homopause, are hydrogen atoms, molecules and helium atoms. For atoms to act as cooling agents, temperatures have to be hot enough to populate excited electronic levels. For H, the energy gap between electronic level 1 and 2 has a temperature equivalent of nearly 25,000K, and for helium atoms it is even higher. The atomic species are thus not involved in cooling the atmospheres of the planets of our Solar System. Molecular hydrogen, H_2, has a vibrational frequency of $\sim$ 4161 cm^{-1}, with an excitation temperature of 1250K, a temperature reached towards the top of Jupiter's thermosphere (Grodent et al. 2001). In principle, therefore, it is a candidate for radiative cooling. But H_2 is a homonuclear diatomic molecule, and thus has neither a permanent dipole moment nor a dipole moment generated by its vibrational mode. H_2 cannot cool by allowed dipole transitions, although it does cool by much less efficient quadrupole transitions.

H_3^+, as has already been discussed, does have strong ro-vibrational bands that run throughout the near- and mid-infrared, with excitation temperatures that run from a few hundred to a few thousand degrees. The near-infrared bands, from the fundamental ν_2 through to the second overtone $3\nu_2$, as well as several of the hot and combination bands, all have Einstein A coefficients of $\sim$ 100 s^{-1} (Dinelli et al. 1992), some seven orders of magnitude greater than the H_2 quadrupole transitions. Moreover, as a polyatomic molecule H_3^+ has many more individual transitions per band than diatomic H_2. All of this means that although H_3^+ densities are considerably lower than those of H_2, it is the ion that is mainly responsible for cooling the upper atmosphere by radiation to space. This cooling effect, now known as the **H_3^+ thermostat**, was first proposed in the early 1990s (Miller et al. 1994). In the jovian auroral regions, it can reach from ten to over a hundred mW m^{-2}. By comparison, the solar EUV absorbed by Jupiter is of the order of 0.3 mW m^{-2} (Yelle & Miller 2004).

Since H_3^+ is produced by the input of energy into the upper atmosphere both by EUV and particle precipitation, it is an interesting question as to how much the heating effects of these processes are balanced by H_3^+ radiation to space. A study showed (Melin et al. 2006), within the accuracies of the jovian observations made by Stallard et al (2001), that H_3^+ cooling pretty much balances the heating due to particle precipitation, even during an intense auroral event. The jovian upper atmosphere is in thermal balance on September 8 (small positive heating), with H_3^+ cooling playing an important role in balancing particle precipitation. On September 11, however, substantially increased energy inputs due to Joule heating and ion drag cannot be offset by the various cooling processes (Table 1).

The H_3^+ thermostat also seems to be applicable for Uranus, but may be less effective for Saturn, which has a thermospheric temperature of $\sim$ 400K, somewhat too low to populate the vibrationally excited levels required for radiation (Melin et al. 2007). More recent Cassini observations of Saturn, however, indicate that the auroral polar re-

Table 1. Heating and cooling rates during an auroral event in Jupiter's northern polar region during September 8 - 11, 1998 [from Melin et al. (2006)].

Heating/cooling term	Sept. 8 W m^{-2}	Sept. 11 W m^{-2}
Joule heating and ion drag	67.0	277.0
Particle precipitation	10.8	12.0
Downward conduction	-0.3	-0.4
$E(H_3^+)$ cooling	-5.1	-10.0
Hydrocarbon cooling	-65.5	-103.3
Net heating rate	**7.4**	**175.3**

gions of Saturn are very variable, and that the temperatures locally may exceed 600K, at which point H_3^+ does become an efficient radiator (Stallard et al. 2008b).

6.2. The H_3^+ Thermostat in Exoplanet Atmospheres

The planets with an orbit very close to their star are subject to intense radiation, which causes considerable heating. Unless that heating can somehow be counteracted, it is inevitable that the atmospheric temperature will rise and the gas expand such that bulk flows may escape from the planet's gravitational field - hydrodynamic escape. The exoplanet HD 209458b, for example, has a mass of 0.685 Jupiter masses, and orbits its G0 star with a semi-major axis of just 0.047AU. In 2003, Vidal-Madjar and coworkers (Vidal-Madjar et al. 2003) reported that HD209458b had an atmosphere that extended for some 2.5 planetary radii, and later, that it was hydrodynamically escaping (Vidal-Madjar et al. 2004). One-dimensional modelling by (Yelle 2004) showed that HD209458b was losing about 4×10^{10} g s^{-1}, a rate that is too slow to affect its overall mass by more than 1% during the lifetime of its central star.

Since HD209458b is clearly very Jupiter-like in mass and in the star it orbits, the question then arises as to at what point during its migration inwards from where it must have formed, its upper atmosphere changed from being Jupiter-like – *i.e.* extending just 0.1 planetary radii beyond the nominal planetary surface, usually set at the 1-bar level for Jupiter – to being extended, "HD209458b-like". Koskinen et al. (2007) approached the problem by asking what would happen if Jupiter were brought gradually closer to the Sun. Using a fully-coupled, three-dimensional global circulation model, they simulated all the physical and chemical processes relevant to a Jupiter-like planet. In particular, they were able to include the energy redistributive properties of winds, the increased production of H_3^+ resulting from increased solar EUV fluxes, the subsequent increased cooling effect of the H_3^+ thermostat, and the dissociation of H_2 molecules, vital to the formation of H_3^+, as a result of increasing temperature.

This modelling produced a rather startling result (Fig. 7). At 0.16 AU from the Sun, the H_3^+ thermostat proved capable of stabilising the upper atmosphere of Jupiter, with infrared cooling offsetting increased heating due to the absorption of high solar EUV fluxes, more than 1,000 greater than those experience by our actual Jupiter. But at 0.14 AU, just 3 million kilometres closer and with an extra 30% EUV insolation compared with 0.16 AU, this thermostat broke down: temperatures rose sufficiently that the thermal dissociation of H_2 became significant; the loss of H_2 resulted in a failure to form sufficient quantities of H_3^+ to maintain cooling. The temperature in the uppermost layers of the atmosphere rose from an average of ~ 3,000 K to over 20,000 K, and the radius of the planet more than doubled. Koskinen et al. (2007) had found that there was a sharp boundary, a stability limit, between the 0.16 AU Jupiter-like planet and the 0.14 AU HD209458b-like planet, a direct consequence of the breakdown of the H_3^+ thermostat.

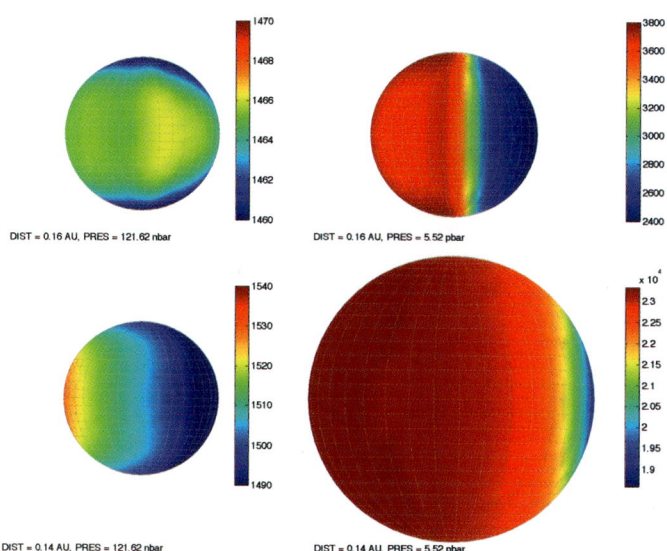

Figure 7. The effect of moving Jupiter closer to the Sun using the University College London, fully-coupled, 3-D global circulation model of the jovian thermosphere and ionosphere. The left-hand column shows conditions at the lower boundary of the thermosphere and the right-hand column, for simulated distances of 0.16 AU (top row) and 0.14 AU (bottom row). H_3^+ cooling keeps the atmosphere stable at 0.16 AU (24 million km from the Sun), but the breakdown of molecular hydrogen, the precursor of H_3^+, at 0.14 AU (21 million km) leads to the failure of the H_3^+ thermostat, very large temperature rises, and hydrodynamic escape of the rapidly expanding atmosphere.

This modelling was carried out using the present-day solar radiation fluxes. But it is well-known that the EUV flux from the Sun that prevails today is much less than in former eras. The "Sun in Time" project (Ribas et al. 2005) makes use of a number of solar analogues that indicate how this flux may have changed over time. Their sample ranges from EK Draconis, at an equivalent age of 0.1 Gyr, to β Hydri at 6.7 Gyr. During this time, the EUV flux falls from 110 times current-day solar to 0.63 of the present value. Each 10-fold increase in age is associated with a ~ 15-fold decrease in EUV flux. Koskinen et al. (2007) calculated that, all other things being equal, the stability limit of a 0.1 Gyr-old Sun would have been at 1.68 AU. This result raises some interesting questions.

Firstly, if all exo-Jupiters were formed at several AU distant from their star (at the ice-formation limit) and then migrated inwards, all those within an equivalent EUV-flux distance of 1.68 AU (about 240 of them) must have passed through an HD209458b-like phase, since migration times for exo-Jupiters are much less than 0.1 Gyr. Secondly, if the increase in EUV flux with decreasing age persists, then at some time around 10 Myr after its formation, the Sun's EUV flux must have been sufficiently intense that Jupiter would also have been inside the stability limit, and would also have suffered a HD209458b-like phase. At some later time in its evolution, then, the H_3^+ thermostat would have been crucial to enabling its atmosphere to condense back to its current extension of about 0.1 planetary radii, from what may have been a hot, hydrodynamically escaping, highly extended state.

6.3. Joule Heating and Ion Drag

So far, we have concentrated on the effect of H_3^+ in cooling the upper atmospheres of giant planets, which results from its ability to radiate strongly throughout the near-infrared region of the spectrum, at the temperatures that prevail in the thermosphere. High-resolution spectroscopy of Jupiter's H_3^+ auroral emission with CSHELL showed that ions were being Doppler-shifted in the line of sight (Rego et al. 1999). The Doppler shifting of very bright lines of H_3^+ can be measured to tenths of a pixel, equivalent to some 300 m s^{-1}. The measured ion wind speeds, v_{ion} were of the order 1–3 km s^{-1}, and follow up studies showed that there was a prevalent (planetary) westward ion drift around the auroral oval, as well as other wind systems poleward of the main oval (Stallard et al. 2001).

This auroral ion drift resulted from powerful electric fields projected from the magnetosphere, in regions where the equatorial plasmasheet could no longer be maintained in corotation with the planet, onto the ionosphere (Hill 1979, 2001; Cowley & Bunce 2001). In the planetary frame of reference, these equatorward electric fields, E_{eqw}, drove currents through the ionosphere, resulting in Joule heating. The observed ion drifts resulted from the Hall effect, given by:

$$v_{ion} = -\mathbf{E_{eqw}} \times \mathbf{B_{aur}}/B_{aur}^2 \qquad (11)$$

where B_{aur} is Jupiter's, near-vertical, magnetic field in the auroral regions, which has a value of some 10^{-3} Tesla (Connerney et al. 1998). Thus, the 1–3 km s^{-1} ion winds implied electric fields of 1–3 V m^{-1}. Millward et al. (2005) used the the time-dependent, fully coupled, three-dimensional, global circulation Jovian Ionospheric Model (JIM) (Achilleos et al. 1998) to show that the ion-neutral collisions generated by the winds along the auroral oval caused the neutral thermosphere in that region to develop a westward drift as well, with the neutrals picking up a fraction, K, of the ion velocity. These authors found that K attained values as large as 0.5 to 0.7 at the peak of the H_3^+ ion density, in the 0.1 μbar pressure region. The drift of the neutral atmosphere with the ions has the effect of lowering Joule heating, but this is partially compensated for as a result of the frictional heating due to ion drag on the thermosphere. Smith et al. (2005) derived an expression for the total electric heating, H_{elec}, due to Joule heating and ion drag:

$$H_{elec} = (1-K)E_{eqw}^2 \Sigma_P \qquad (12)$$

where Σ_P is the Pedersen conductivity.

JIM simulations showed that H_3^+ was produced at just the right altitude to generate large values of Σ_P, between 1 and 10 mho (Millward et al. 2002). This model result has now been confirmed by observations of the H_3^+ vertical profile as a function of altitude, made using NIRSPEC on the 10-m Keck II Telescope (Lystrup et al. 2008). The high conductivities result in the total electric heating generated in the auroral oval region of Jupiter being significantly greater than 10^{14} W, several hundred times greater than the energy absorbed from EUV insolation planetwide. Measurements of ion drifts in Saturn (Stallard et al. 2004) indicate that several times 10^{12} W of energy is generated in its auroral/polar regions as a result of Joule heating and ion drift (Miller et al. 2006). This sum is, again, well over an order of magnitude greater than the heating due to EUV insolation. So far, ion winds have not been detected for Uranus. This may be as a result of its highly complex magnetosphere and its greater distance from Earth.

These considerations show that as well as cooling planets, H_3^+ ions couple with electric fields, generated by magnetosphere-ionosphere interactions, to produce a great deal of heating. Indeed, this heating may help to explain why the upper atmospheres of all of the Solar System's giant planets are much hotter than can be explained by the absorption of solar EUV radiation (Yelle & Miller 2004).

7. Conclusions : Future of H_3^+ Search

All the elements discussed in Sect. 6 demonstrates that the possibility of detecting H_3^+ in extrasolar giant planets would provide a fundamental probe of their upper atmosphere, of their thermal balance, and possibly of their magnetic field. At present, we have no knowledge of whether the exoplanets so far discovered are magnetised, although it would be surprising if none of them were, given that five of the Solar System's eight planets have large permanent magnetic fields. Our lack of knowledge of exoplanetary magnetic field conditions, of their magnetospheres and thus, magnetosphere-ionosphere coupling, means that we cannot tell if H_3^+ will also have a role as a heating agent for Jupiter-like exoplanets, too. Again, however, it would be a surprise if it did not, given what we know about Jupiter, Saturn and Uranus, which has been reviewed above.

The best current upper limit of detection of H_3^+ in giant exoplanets reported by Laughlin et al (2008) brings to discussion the feasibility of using ground-based high-resolution spectroscopy for this purpose. It brings also the question of the ubiquity of H_3^+ in these atmospheres. One important parameter is the distance between the parent star and its planet. As seen in Section 6.2 to work the H_3^+ thermostat requires the giant planet to be at a distance beyond a strict limit which depends on the characteristics of the star. Too close, too much molecular hydrogen is dissociated into atomic hydrogen and H_3^+ cannot form. In addition, a strong stellar radiation, particularly from a very young star, can induce a rapid thermal evaporation of the giant planet atmosphere. This phenomenon may lead to a population of Neptune-mass planets and super-Earth which becomes to be detected (Bonfils et al. 2007), which might be hydrogen-poor planets with mainly a dense core remnant. Too far, the EUV flux is not high enough and the only chance is a magnetised giant planet. Hence, there is a problem of strategy of observation. The most favorable targets might be several Jupiter's mass planet not much closer then 0.1 AU from their parent star.

On the other hand, at least a gain of a factor 10 in sensitivity is needed to improve the chance of detection and possibly, with a higher spectral resolution than with an instrument like CSHELL. This gain may have to wait for high-resolution infrared spectrometers behind an ELT. A promising technique, already used for the first detections of molecules in extrasolar planet atmospheres (this conference) is the *occultation spectroscopy*, making use of primary and secondary transits for a difference of the two spectra. With the future missions more transiting planets should be known, providing a larger sample of targets, and therefore, offering more favorable conditions to detect H_3^+ in the these planets.

References

Abel, B., & Thorne, R.M. 2003, Icarus, 166, 311
Achilleos, N., Miller, S., Aylward, A.D., Mueller-Wodarg, I., & Rees, D. 1998, J. Geophys. Rev., 103, 112
Alijah, A., & Varandas, A.J.C. 2006, Phil. Trans. Royal Soc. London, 364, 2889
Baron, R., Joseph, R.D., Owen, T.. Tennyson, J., Miller, S., & Ballester, G.E. 1991, Nat, 353, 539
Bawendi, M.G., Rhefuss, B.D., & Oka, T. 1990, J. Chem. Phys., 93, 6200
Bonfils, X., Mayor, M., Delfosse, X., et al. 2007, A&A, 474, 293
Brittain, S., & Rettig, T.W. 2002, Nat, 418, 57
Broadfoot, A.L., Belton, M.J., Takacs, P.Z., et al. 1979, Science, 204, 979
Cowley, S.W.H., & Bunce, E.J. 2001, Planet. Space Sci., 49, 1067
Connerney, J.E.P., Acuna, M.H., Ness, N.F., & Satoh, T. 1998, J. Geophys. Res., 103, 11929
Dinelli, B.M., Miller, S., & Tennyson, J. 1992, J. Mol. Spec., 153, 718
Drossart, P., Maillard, J.P., Caldwell, J. et al. 1989, Nat, 340, 539
Drossart, P., Prangé, R., & Maillard, J.P. 1992, Icarus, 97, 10
Encrenaz, T., Drossart, P., Orton, G., Feuchtgruber, H., Lellouch, E., & Atreya, S.K. 2003, Planet Sp. Sc., 51, 1013
Geballe, T.R., Jagod, M.F., & Oka, T. 1993, ApJ, 408, L1009
Geballe, T.R., & Oka, T. 1996, Nat, 384, 334
Goto, M., Geballe, T.R., McCall, B.J., Usuda, T., Suto, H., Terada, H., Kobayashi, N., & Oka, T.(2005, ApJ, 629, 865
Grodent, D., Waite, J.H. Jr., & Gérard, J.C. 2001, J. Geophys. Res, 106, 12933
Herbst, E., & Klemperer, W. 1973, ApJ, 185, 505
Hill, T.W. 1979, J. Geophys. Res., 84, 6554
Hill, T.W. 2001, J. Geophys. Res., 106, 8101
Hirschfelder, J.O. 1938, J. Chem. Phys., 6, 795
Kim, S., Caldwell, J., Rivolo, A.R., & Wagener, R. 1985, Icarus, 64, 233
Kim, S.J., Drossart, P., Caldwell, J., & Maillard, J.P. 1990, Icarus, 84, 54
Koskinen, T.T., Aylward A.D., & Miller, S. 2007, Nat, 450, 845
Lam, H.A., Achilleos, N., Miller, S., Tennyson, J., Trafton, L.M., Geballe, T.R., & Ballester, G. 1997 Icarus, 127, 379
Laughlin, L., Troutman, M.R., Brittain, S., & Rettig, T.W. 2008, BAAS, 212, 10.11
Lee, S.S., Ventrudo, B.F., Cassidy, D.T., Oka, T., Miller, S., & Tennyson, J. 1991, J. Mol. Spec., 145, 222
Lindsay, C.M., & McCall, B.J. 2001, J. Mol. Spect., 210, 60, http://h3plus.uiuc.edu/
Lystrup, M.B., Miller, S., Dello Russo, N., Vervack, R.J., Jr., & Stallard, T. 2008, ApJ, 677, 790
Majewski, W.A., Feldman, P. A., Watson, J.K.G., Miller, S., & Tennyson, J. 1989, ApJ, 347, 51
Maillard, J.P., & Michel, G. 1982 in Instrumentation for Astronomy with Large Optical Telescopes, ed. C.M. Humphries (Dordrecht: Reidel), 213
Maillard, J.P., Drossart, P., Watson, J.K.G., Kim, S.J., & Caldwell, J. 1990, ApJ, 363, L37
Maillard, J.P., Lellouch, E., Waite, H., Jr., Bézard, B., Drossart, P., Mandin, J.Y., & Dana, V. 1999, BAAS, 31, 31.07
Marten, A., de Bergh, C., Owen, T., Gautier, D., Maillard, J.P., Drossart, P., Lutz, B.L., & Orton, G.S. 1994, Planet. Sp. Sc., 42, 391
Martin, D.W., McDaniel, E.W., & Meeks, M.L. 1961, ApJ, 134, 1012

Mayor, M., & Queloz, D. 1995, Nat, 378, 355
McCall, B.J. 2000, Phil. Trans. Royal Soc. London, 358, 2385
Melin, H., Miller, S., Stallard, T., & Grodent, D. 2005, Icarus, 178, 97
Melin, H., Miller, S., Stallard, T., & Grodent, D. 2006, Icarus, 181, 256
Melin, H., Miller, S., Stallard, T., Trafton, L.M., & Geballe, T.R. 2007, Icarus, 186, 234
Meyer, W., Botschwina, P., & Burton, P. 1986, J. Chem. Phys., 84, 891
Miller, S., & Tennyson, J. 1987, J. Mol. Spec., 128, 183
Miller, S., Tennyson, J., & Sutcliffe, B.T. 1989, Mol. Phys., 66, 429
Miller, S., Tennyson, J., & Sutcliffe, B.T. 1990, J. Mol. Spec., 141, 104
Miller, S., Lam, H.A., & Tennyson, J. 1994, Can. J. Phys., 72, 760
Miller, S., Achilleos, N., Ballester, G.E., Lam, H.A., Tennyson, J., Geballe, T.R., & Trafton, L.M. 1997, Icarus, 130, 57
Miller, S., Achilleos, N., Ballester, G.E., Geballe, T., Joseph, R.D., Prangé, R., Rego, D., Stallard, T., Tennyson, J., Trafton, L.M., & Waite, J.H. 2000, Phil. Trans. Royal Soc. London, 358, 2485
Miller, S., Stallard, T., Smith, C., Millward, G., Melin, H., Lystrup, M., & Aylward, A.D. 2006, Phil. Trans. Royal Soc. London, 364, 3121
Millward, G., Miller, S., Stallard, T., Achilleos, N., & Aylward, A.D. 2002, Icarus, 160, 75
Millward, G., Miller, S., Stallard, T., Achilleos, N., & Aylward, A.D. 2005, Icarus, 173, 200
Oka, T. 1980, Phys. Rev. Lett., 45, 531
Oka, T. 1992, Rev. Mod. Phys., 64, 1141
Pan, F.S., & Oka, T. 1986, ApJ, 305, 518
Rank, D.M., Townes, C.H., & Welch, W.J. 1971, Science, 174, 1083
Raynaud, E., Lellouch, E., Maillard, J.-P., Gladstone, G.R., Waite, J.H., Bézard, B., Drossart, P. & Fouchet, T. 2004, Icarus, 171, 133
Rego, D., Achilleos, N., Stallard, T., Miller, S., Prangé, R., Dougherty, M., & Joseph, R.D. 1999, Nat, 399, 121
Rego, D., Miller, S., Achilleos, N., Stallard, T.S., Prange, R., Dougherty, M., & Joseph, R.D. 2000, Icarus, 147, 366
Ribas, I., Guinan, E.F., Gudel, M., & Audard, M. 2005, ApJ, 622, 680
Shkolnik, E., Gaidos, E., & Moskovitz, N. 2006, AJ, 132, 1267
Smith, C., Miller, S., & Aylward, A.D. 2005, Ann. Geophys., 23, 1943
Stallard, T., Miller, S., Millward, G., & Joseph, R.D. 2001, Icarus, 154, 475
Stallard, T., Miller, S., Trafton, L.M., Geballe, T.R., &. Joseph, R.D. 2004 Icarus, 167, 204
Stallard, T., Lystrup, M., & Miller, S. 2008, ApJ, 675, L117
Stallard, T., Miller, S., Lystrup, M. et al. 2008, Nat, 456, 214
Sutcliffe, B.T., Tennyson, J., & Miller, S. 1987, Theor. Chim. Acta, 72, 265
Thomson, J.J. 1911, Phil. Mag., 21, 225
Trafton, L.M., Geballe, T.R., Miller, S., Tennyson, J., & Ballester, G.E. 1993, ApJ, 405, 761
Trafton, L.M., Miller, S., Geballe, T.R., Tennyson, J., & Ballester, G.E. 1999, AJ, 524, 1059
Vidal-Madjar, A., Lecavelier des Etangs, A., Désert, J.M., Ballester, G.E., Ferlet, R., Hébrard, G., & Mayor, M. 2003, Nat, 422, 143
Vidal-Madjar, A., Désert, J.M., Lecavelier des Etangs, A., Hébrard, G., Ballester, G.E., Ehrenreich, D., Ferlet, R., McConnell, J.C., Mayor, M., & Parkinson, C.D. 2004, ApJ, 604, L69
Xu, L.W., Rosslein, M., Gabrys, C.M., & Oka, T. 1992, J. Mol. Spec., 153, 726
Yelle, R.V. 2004, Icarus, 170, 508
Yelle, R., & Miller, S. 2004, in Jupiter: The planet, satellites and magnetosphere, ed. F., Bagenal, T.E., Dowling, & W.B., McKinnon, (Cambridge University Press), 185

Wesley Traub and James Cho

Part II

Exoplanet Atmospheres from Transit Observations
(Chairs : Gautam Vasisht and Giovanna Tinetti)

Lessons Learned from Ground-based Transmission Spectroscopy of Extrasolar Planets

I.A.G. Snellen, S. Albrecht, E.J.W. de Mooij, and R.S. Le Poole

Leiden Observatory, Leiden University, Postbus 9513, 2300 RA, Leiden, The Netherlands

Abstract. After many unsuccessful attempts, ground-based transmission spectroscopy has finally yielded its first detections of planetary atmospheres. Redfield et al. (2008) have presented successful observations of sodium in HD189733b using the Hobby-Eberly Telescope, and our group have detected the same chemical element in HD209458b using archival Subaru data (Snellen et al. 2008). In this conference contribution we first argue that in the light of the earlier successes with the Hubble Space Telescope, it would have been very surprising if ground-based transmission spectroscopy had remained fruitless. Secondly, we discuss the scientific capabilities unique to high dispersion spectroscopy from the ground, and we conclude with giving some advice on how to perform successful future observations.

1. Transmission Spectroscopy of Extrasolar Planets

With transmission spectroscopy, the depth of a planet transit is measured as function of wavelength. At certain wavelengths a transit will be slightly deeper due to absorption in the planet's atmosphere, and in this way its atmospheric constituencies can be determined. In the optical transmission spectra of hot Jupiters, the strongest of these absorption features was predicted to come from the sodium D lines at 5889Å and 5896Å (Brown 2001; Seager & Sasselov 2000). Indeed, Charbonneau et al. (2002) detected extra absorption due to sodium from the transiting exoplanet HD209458b, at a level of $0.023 \pm 0.006\%$ in a 12Å wide band, using the STIS spectrograph on the Hubble Space Telescope (HST). Subsequently, several other chemical elements and molecules have been detected in one or both of the two large and bright transiting systems, HD209458 and HD189733, using either the HST or the Spitzer Space Telescope. Particularly strong absorption features have been detected in HD209458b from Hydrogen, Carbon, and Oxygen at a level of 5-15%, thought to be originating from an evaporating exosphere (Vidal-Madjar et al. 2003; 2004). In addition, recent observations have revealed absorption signatures from water, methane, and CO (Tinetti et al. 2007; Swain et al. 2008).

Before the recent detections of sodium in HD189733b and HD209458b by Redfield et al. (2008) and our group (Snellen et al. 2008) respectively, ground-based transmission spectroscopy had not been a great success. Typically, upper limits to planetary absorption from sodium D of 0.1-1% had been presented (e.g. Moutou et al. 2001; Snellen 2004; Narita et al. 2005), and ground-based transmission spectroscopy had more or less been given up as a possible tool to learn more about extrasolar planet atmospheres.

1.1. Why Ground-based Transmission Spectroscopy Should Work

One could say that it is quite surprising that ground-based transmission spectroscopy did not deliver results for so long. Although the HST result by Charbonneau et al. (2002) meant that the relative transit signal was expected to be very small (2×10^{-4}), this signal was measured over a 12Å wide band. For a reasonable double-line planetary absorption profile this should easily be detectable with an Echelle spectrograph on a 10m class telescope. Simply speaking, assuming that the actual atmospheric absorption consists of two 1Å wide features, the 2×10^{-4} absorption measured with the HST should correspond to a 6× higher level of absorption (1.2×10^{-3}) in a 1 Å passband. As we will argue in section 3, the narrow planetary absorption needs to be even higher because it absorbs the star-light at a wavelength were the stellar sodium absorption has already eaten away a large fraction of the flux. Echelle spectrographs on ten-meter class telescopes deliver spectra with typically signal-to-noise ratios (SNR) of a few hundred per resolution element, and an SNR of one thousand or more per Angström, within only a few minutes of exposure time for the brightest transiting exoplanet host stars. Hence, over the whole transit, this should correspond to a detection of the double 1Å wide sodium features in HD209458b at a 5-10σ level, meaning that the previous unsuccessful ground-based work must have been strongly affected by systematic effects, at a level at least 5 times higher than the photon noise level.

2. Re-analysis of Subaru HDS Data on HD 209458b

We searched the databases of large telescopes for the best available data set that covered a transit of HD209458b, which would allow us to assess the reason(s) for not detecting the planetary sodium absorption. In addition, with a fresh look at the data, we could possibly manage to reduce systematic effects, hopefully revealing the sodium absorption after all. We chose to re-analyse the data from the High Dispersion Spectrograph (HDS) on the Subaru Telescope, as described by Narita et al. (2005). The data consists of 18 spectra taken during transit, and 12 spectra outside transit, at a spectral resolution of ~45,000, each with a SNR of about 300-450 per pixel (SNR ~3,000/Å).

2.1. What Was New About Our Data Analysis?

For a detailed description of the data analysis see Snellen et al. (2008). While for most of our data analysis we followed the original procedure as outlined by Narita et al. (2005), we changed the data reduction on a few crucial points:

1. First of all, in Narita et al. strong ripples show up at the position of the stellar sodium lines when the differences between the individual spectra and an averaged template are plotted. We think these were caused by small, random wavelength offsets between the individual spectra, induced by telescope pointing errors causing the star to change position in the slit. We therefore first cross-correlated and aligned all spectra to make the stellar absorption features match in wavelength.

2. Secondly, we found that the depth of the stellar absorption lines in the individual frames were strongly correlated with the count-level in the stellar continuum, c.f. the more flux present in the spectra, the deeper the lines. After eliminating several other possibilities, we attributed this to a non-linearity effect in the CCD, a dependence of the conversion factor between electrons and counts on the total

count level, for which the higher the count level in a pixel the smaller this conversion factor. In this way, the higher the continuum count level, the more the continuum level is overestimated and the lines will appear deeper. We corrected for this effect in an ad hoc way by de-correlating the measured strength of the sodium absorption with the normalized continuum level. Note that Narita et al (2009, this volume) have recently measured the non-linearity of the HDS CCDs to be in agreement with this effect.

3. A big challenge in ground-based transmission spectroscopy, in particular for measuring sodium absorption, is the removal of contamination from telluric water and telluric sodium. In most previous analyses, this contamination is dealt with by observing a fast-rotating A-star which spectrum is scaled and removed from the target spectra. However, our experience is that this does not work very well, likely because stellar spectra are also influenced by interstellar sodium absorption which is different for every star, and which hampers a correct removal of the telluric contamination. We therefore took a very different approach. Instead we de-correlated the measured sodium absorption with the equivalent width of some strong telluric H_2O lines outside the region of interest. Note that for the Subaru data, the equivalent widths of these H_2O lines follow the variation in airmass very closely, indicating it was a perfectly photometric night.

2.2. Detection of Sodium

The level of sodium absorption was measured in all spectra in passbands centered around the two stellar sodium absorption lines, with widths of 3.0, 1.5 and 0.75 Å, and subsequently de-correlated to account for the non-linearity and telluric effects (see above). The sodium absorption due to the planet's atmosphere is detected at $>5\sigma$, at a level of 0.056±0.007%, 0.070±0.011%, and 0.135±0.017% in the three bandwidths respectively (see top panel of Figure 1). The Subaru observations are fully consistent with the STIS/HST detection. The measurements in the two narrowest bands indicate that some signal is being resolved (see bottom panel of Fig. 1), meaning that the width of the planetary absorption feature is about ~1Å (quite similar to the width of the stellar absorption actually).

2.3. Comparison with the Sodium Detection in HD 189733b

We also compared our sodium detection with that observed in HD189733b. Redfield et al. (2008), who measured this absorption at a level of 0.067±0.021% in a passband of 12Å, argue that the sodium absorption in the atmosphere of HD189733b is 3× higher than that of HD209458b, indicating that their atmospheres are very different. However, a possible pitfall is that in their analysis they divided their average in-transit spectrum by the average out-transit spectrum before integration over the 12Å passband ($\int_{\Delta\lambda} F_{in}/F_{out}$), instead of first integrating over the passband before dividing the in-transit by the out-transit data ($\int_{\Delta\lambda} F_{in} / \int_{\Delta\lambda} F_{out}$). In the first case, as utilized by Redfield et al. the relative weights assigned to the pixels in the heart of the stellar sodium lines are more than an order of magnitude too high. This means that the sodium absorption is actually only about a factor 2 stronger than in HD209458b, implying that the scale heights in sodium are very similar for the two planets, because the HD189733b transit is also almost a factor 2 deeper than that of HD209458b.

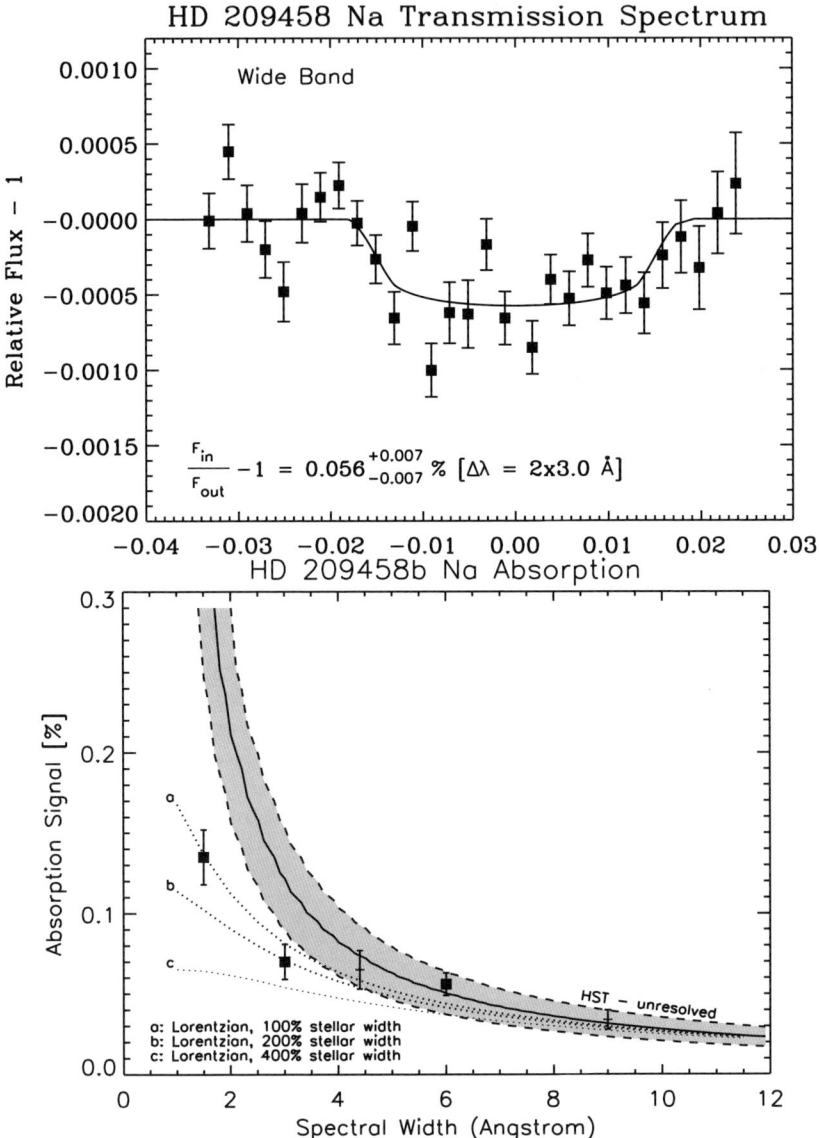

Figure 1. Upper panel: Transit photometry of the Na D doublet as measured in a 3.0Å band with Subaru (Snellen et al. 2008). Lower panel: Measurement of the Na D absorption in three different bands, compared to STIS/HST observations as analysed by Charbonneau et al. (2002) and Sing et al. (2008).

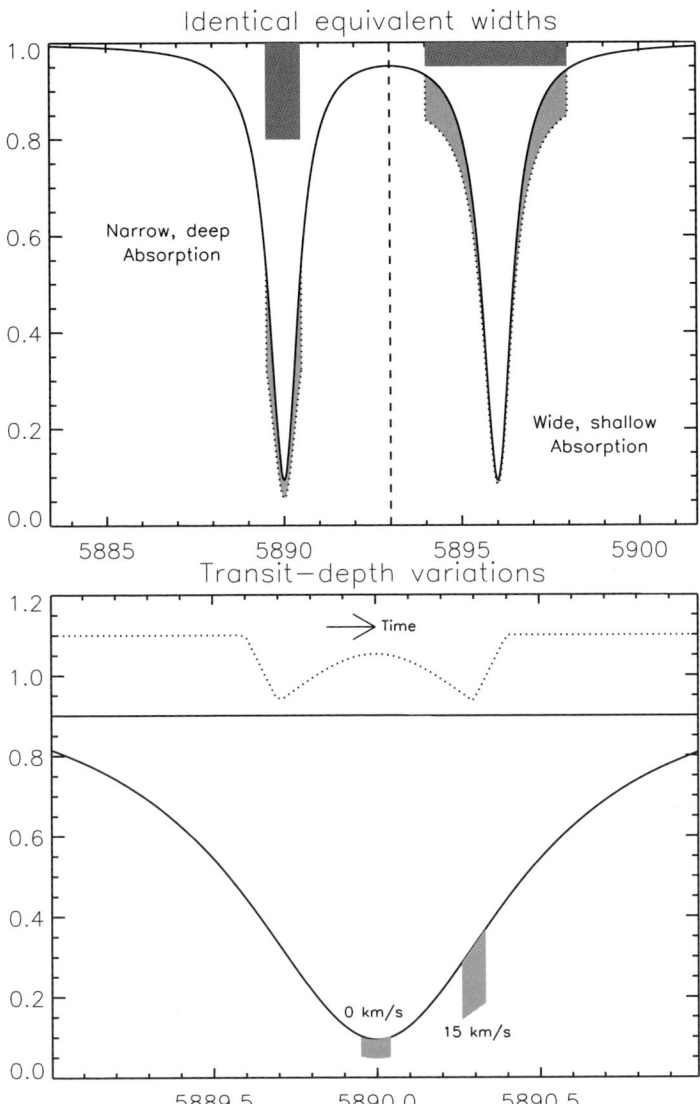

Figure 2. Upper panel: A schematic representation of two absorption features of similar strength but different widths. The wider feature (on the right) produces a much stronger absorption because it also covers the stellar spectrum outside the core of the stellar sodium absorption line. Lower panel: Schematic drawing to show that during a transit the measured absorption strength can also vary due to the fact that the relative position of the planetary absorption changes because of the orbital velocity of the planet.

Redfield et al. also measure a possible velocity offset for the planetary absorption of ~38 km/sec blueward with respect to the star. We can just note that we do not see such an offset for HD209458b. If this velocity shift is real, it is rather puzzling how it could be produced since it is about an order of magnitude higher than the expected sound speed in the upper layer of the planet's atmosphere.

3. The Unique Aspects of Ground-based Transmission Spectroscopy

Unfortunately, it seems that the Hubble Space Telescope will not be usable for transmission spectroscopy anymore. Since the James Webb Space Telescope will not be operating at wavelengths of the sodium absorption, optical transmission spectroscopy will be very limited from space, also for the foreseeable future. Already for this reason it is important to further explore the limits and possibilities of ground-based spectroscopy. Another point is that ground-based observations allow the use of Echelle spectrographs, providing spectral resolution unattainable from space. As we show with our work, this can provide interesting information on the spectral shape of the planetary absorption, which should be fed back into exoplanet atmosphere models (see e.g. Sing et al. 2008). Ground-based transmission spectroscopy could also be very fruitful in the near-infrared, where high spectral resolution can resolve molecular absorption bands and can be used to combine the signal of tens to hundreds of individual molecular absorption lines (from e.g. water, CO, or methane).

4. Some Notes and Advice for Future Observations

We would like to conclude with some notes on ground-based transmission spectroscopy, and some advice for future observations, based on our experiences:

Note 1: Width of sodium absorption influences the measured depth. The measurement of sodium absorption in a wide band cannot be translated directly to an absorption scale height, without knowing the spectral width of the feature. if it is very narrow, most absorption occurs in the heart of the stellar absorption line where the stellar photon flux is an order of magnitude lower than in the spectral continuum. It would require a much stronger level of absorption to reach the same broad-band level than for a wider absorption feature (see upper panel Fig. 2).

Note 2: Planet velocity influences measured depth. The measured planetary sodium absorption can vary over the duration of the transit, because of the change in radial velocity of the planet. For HD209458b, this moves the center of the sodium absorption from −15 km/s to +15km/sec over the time scale of the transit. This means that at the beginning and end of the transit the planetary atmospheric absorption is located blueward and redward with respect to the stellar absorption lines, influencing the measured absorption as in the case above (see bottom panel Fig 2).

Advice 1: Make use of the whole observing night. Although transits last typically for not more than 2 to 3 hours, in-transit data is as valuable as out-transit data. Therefore make sure to observe the target for as long as possible. It is also very useful to observe over an as wide as possible airmass range, needed to de-correlate the data. Ideally the transit does not occur during meridian passage avoiding low-airmass data to be associated with in-transit observations, and high-airmass with out-transit data.

Advice 2: Avoid observing calibrator stars. Our experiences are that these calibrations are not sufficient to remove telluric contaminations, and therefore should be avoided to save time and to avoid breaks and offsets in the observing sequence.

Advice 3: Correct for telluric contamination by de-correlation with strong H_2O lines (see section 2.1).

Advice 4: Limit analysis to relatively wide spectral bands, a) because variations in seeing can result in variations in spectral resolutions (for a slit spectrograph), b) the shape of the stellar lines vary during the transit due to the Rossiter--McLaughlin effect, and c) the relative radial velocity of the planet could otherwise make the absorption signal (partly) move outside the passband.

Acknowledgments. We wish to thank the organisers for this very interesting workshop. The data in this contribution are collected at the Subaru telescope and obtained from the SMOKA, which is operated by the Astronomy Data center, National Astronomical Observatory of Japan.

References

Brown, T. M. 2001, ApJ, 553, 1006
Charbonneau D., Brown T.M., Noyes R.W., Gilligand R.L., 2002, ApJ568, 377
Moutou, C., et al. 2001, A&A, 371, 260
Narita, N., et al. 2005, PASJ, 57, 471 (NAR05)
Redfield, S., Endl, M., Cochran, W. D., & Koesterke, L. 2008, ApJ, 673, L87
Seager, S., & Sasselov, D. D. 2000, ApJ, 537, 916
Sing, D. K., et al. 2008, ApJ, 686, 658
Snellen, I. A. G. 2004, MNRAS, 353, L1
Snellen, I.A.G., Albrecht, S., de Mooij, E.,J.W., & Le Poole, R. S. 2008, A&A, 487, 357
Swain, M. R., Vasisht, G., & Tinetti, G. 2008, Nat, 452, 329
Tinetti, G., et al. 2007, Nat, 448, 169
Vidal-Madjar, et al. 2003, Nat, 422, 143
Vidal-Madjar, A., et al. 2004, ApJ, 604, L69

Giovanna Tinetti, Mark Swain, and Jean-Phillipe Beaulieu opening a session.

The Search for Exomoons

D. M. Kipping

Dept. of Physics & Astronomy, University College London, Gower Street, London WC1E 6BT, UK

Abstract. With exoplanet detections becoming routine, astronomers are now vying to characterise these alien worlds. As well as detecting the atmospheres of these exoplanets, part of the characterisation process will undoubtedly involve the search for extrasolar moons. In this work, we explore the motivations for searching for exomoons, review some of the previously proposed detection techniques and finally introduce transit duration variation (TDV) as a proposed search method. We find that these techniques could easily detect Earth-mass exomoons with current instruments and potentially down to Galilean mass moons with future space missions like Kepler.

1. Introduction

In the past decade, the number of known exoplanet systems has soared from a handful to well over 300, and counting. The sentiment of many astronomers is that the era of characterisation is upon us and it is time to endeavour to characterise these distant, alien worlds. Whilst significant advances have been made in the field of understanding the atmospheres of these worlds, for example with the detection of water vapour in the atmosphere of an extrasolar planet by Tinetti et al. (2007), many other planetary attributes remain a mystery.

One example is the question as to whether exoplanets frequently host satellites, as many of the planets of own solar system do. At the time of writing, no detection of an exomoon has ever been made but this is most probably due to a selection effect of the insensitivity of current techniques. In this work, we explore the motivation for undergoing such a search and present a new detection method based on transit duration variation (TDV).

2. Motivation

2.1. A Novel Detection and Proof of Principle

Although never a sufficient goal in itself, one obvious motivation for attempting to detect an exomoon is because it has never been done before. With no detections ever made, it is difficult to say what scientific developments will be made as a result of such a detection. For example, before the detection of any exoplanets, there seemed to be no reason to doubt the Copernican Principle that the solar system was typical in the galaxy, but the host of planetary detections seem to indicate that this is not the case, as supported by Thommes et al. (2008).

A detection would also serve as a proof of principle that our methods are valid. The theory of exomoon detection begun with Sartoretti & Schneider (1999) proposing the use of transit timing. This theory was developed further by Deeg (2002), Simon et al. (2007) and most recently by Kipping (2009).

2.2. A New Class of Objects

The moons of our own solar system are either formed of rock or ice or a mixture of the two. With most known exoplanets being gas giants (see http://exoplanet.eu), there is an obvious desideratum to expand the number of known rocky or icy bodies and moons could provide such objects. Although this expectation is based on the Copernican Principle, this has already been seen to be a quite dubious supposition in exoplanet science and perhaps the most interesting science would come from the unexpected discovery or moons of an entirely different composition, for example the hypothetical carbon-class planet proposed by Seager et al. (2007).

2.3. Implications for Astrobiology

Although somewhat more speculative, it would seem reasonable to say that exomoons could be objects of intense interest for astrobiologists. Some of the most promising candidates for extraterrestrial habitable locations within our own solar system are the moons of gas giants, for example Europa as discussed by Reynolds et al. (1983). In support of this view, Scharf (2006) suggested exomoons could support habitable environments even outside of the conventional habitable zone due to tidal heating from the host planet.

Another interesting question is how important the Moon was to life on the Earth. Several authors, such as Lathe (2004), have suggested the presence of our Moon makes the Earth particularly suited for complex life. Thus even if we detect an Earth-like planet in the future, the search for a moon around such a body may be critical in assessing its habitability.

2.4. Planetary Formation Theory

Finally, we consider that the detection of exomoons would play an important role in the field of planetary formation theory. It may allow several theories of the stability of exomoons to be tested, for example that of Domingos et al. (2006) and Barnes & O'Brien (2002). It would also allow a test of whether most satellite systems agree with the mass-scaling relation proposed by Canup & Ward (2007).

3. Possible Methods of Detection

3.1. Direct Imaging

We first consider direct imaging as a possible exomoon detection technique. Even if a coronographic or interferometric technique can be utilised to extinguish the star light, the moon light must still compete with the planetary light. The most significant hurdle is therefore likely to be achieving an angular resolution sufficient to resolve the two objects. At a distance of 10pc, the Moon-Earth separation is 0.5 mas whereas the current best interferometric precision in exoplanet astronomy is ~ 25 mas, from Baines et al. (2008). Therefore, one would need at least two coronographic telescopes combined with an interferometer in order to image such systems and there exists no such design in planning. Ergo, we conclude that direct imaging is not a feasible detection method at present.

3.2. Radial Velocity

Radial velocity purely measures the motion of the host star and for a planet-moon system this would appear ostensibly identical to that of a slightly more massive planet. Cabrera & Schneider (2007) argue that if the emitted light of an exoplanet could be measured, then doppler spectroscopy of the planet itself could be achieved which would allow for satellite detection. However, achieving high spectral resolutions of light from a planet alone remains a challenging prospect.

3.3. Pulsar Timing

Lewis et al. (2008) were able to show that the moons of pulsar planets could be detected through time-of-arrival analysis. The authors show that the technique could be sensitive down to moons of mass as low as 5% of that of the host planet. Perhaps it would not be surprising if the first exomoon was found through this method given that the first exoplanet was discovered around a pulsar star by Wolszczan & Frail (2007).

3.4. Occultations

For a sufficiently large moon, the satellite could induce a dip in the starlight. Using this technique, Brown et al. (2001) were able to set an upper limit of 1.2 $R_\oplus$ for an orbiting exomoon around HD209458b. The HST STIS photometry remains at the level of the best achievable photometry to date and thus the sensitivity to such moons has not changed. The occultation method does suffer from the problem that you require a 'lucky' detection. The exomoon could well be hiding in front of behind the planet the moment of transit. Another way of putting this is that the modal position of the moon's lightcurve perfectly coincides with the planetary lightcurve, as pointed out by Cabrera & Schneider (2007). This means one would likely require multiple transits in order to see just one clean moon transit. After this one moon transit had been observed, the whole process would need to repeated several times in order to show that the event was not a statistical fluke or some form of noise.

3.5. Microlensing

The final detection method we will mention here is the microlensing technique proposed by Han (2008). Microlensing is more sensitive to distant planets than the transit method would be, due to the constraint of the Einstein radius of the planet. Microlensing could detect a large number of moons beyond the snow-line of stellar systems and thus aid in the statistical side of exomoon detections. However, microlensed planets and moons could not be followed up due to their great distances from the Earth and thus characterising these moons would be unlikely.

4. Transit Timing Effects

4.1. Transit Time Variation (TTV)

Sartoretti & Schneider (1999) pointed out that an exomoon should induce reflex motion in the host planet as it orbits a star. This reflex motion means that the moment of mid-transit can vary on the order of seconds to minutes and thus could be used as a detection technique. Unfortunately, transit timing variation can also be induced by a multitude of phenomenon, including general relativistic precession of the orbit, gravitational influence of other planets in the system or companion stars, torques due to spin-induced quadrupole moment of the star, tidal deformations of both the star and the planet, stellar peculiar motion and parallax effects (Miralda-Escude (2002), Heyl & Gladman (2007), Ribas et al. (2008), Pál & Kocsis (2008), Jordan & Bakos (2008), Rafikov (2008) and Scharf (2007)). Due to the rich plethora of phenomenon that could cause TTV, claiming any observed signal was indeed due to a moon would be extremely challenging.

Another critical problem with TTV from an exomoon is that the period of an exomoon must always be much smaller than the period of the host exoplanet, due to stability arguments outlined by Kipping (2009). As a result, the observed signal will always be under-sampled and have a frequency above the Nyquist rate. This means that it would be impossible to deduce an accurate period for the exomoon; only a list of possible harmonic frequencies could be derived. The knock-on effect is that since the TTV amplitude is proportional to mass of the exomoon (M_S) multiplied by the orbital distance of the moon (a_S), then it is also impossible to derive the mass of the exomoon through TTV.

4.2. Transit Duration Variation (TDV)

Kipping (2009) was able to show that both these problems with TTV could be resolved if one uses transit duration variation (TDV) in combination with TTV. In regard to the degeneracy problem, TDV is predicted to be 90 degrees out-of-phase with TTV which offers a unique exomoon signature. This signature should be far more reliable than using TTV alone.

TDV also has a different proportionality than TTV. Kipping (2009) showed that TDV should be proportional to $M_S a_S^{-1/2}$ and thus the ratio of TDV to TTV, η, allows one to directly obtain the period of the exomooon. This can then be compared to the set of possible harmonic frequencies to further constrain the period. After the period has been determined, it is trivial to derive the exomoon mass.

5. Feasibility

Despite the advances in the theory of detection, one must question how feasible such an undertaking really is. The question of feasibility really falls into two key requirements i) exomoons exist ii) they are detectable.

In regard to the first question, it would seem natural to assume that long period gas giants frequently harbour satellite systems, but all of the transiting planets discovered so far are essentially gas giants on extremely close-in orbits. Barnes & O'Brien (2002) tackled this question and found that moons of essentially any mass are permitted for planets on distant orbits but strong dynamical constraints limit the possible mass of an exomoon for a given age. For example, if GJ436b had a moon that was 6 Gyr old, the maximum allowed mass would roughly the same as that of Enceladus. However, this calculation does not account for planetary orbital eccentricity, the effect of multiple resonant moons or possible captured moons later in the life of the planet. To further spice this debate, the author would like to point out that close-in exoplanets were not predicted before their discovery and many more surprises may await the dynamicists. Nevertheless, as transit searches progress, we can expect a more promising candidate list to emerge.

The second consideration is whether such exomoons would be detectable, even if they did exist. Kipping (2009) evaluated the timing effects for several of the best transiting candidates in the case of a $1 M_\oplus$ exomoon and found that the timing signal of Neptune-mass host planets should be between 10 and 20 seconds. With some current ground based instruments reaching 7 seconds, for example Johnson et al. (2008), and the Kepler mission expected to achieve sub-second accuracy (see http://kepler.nasa.gov/sci/) then we are already at the level where we can start searching for large exomoons. As methods improve, we can expect our sensitivity to drop down to Galilean mass moons. In figure 1, we show the sensitivity limits for detecting super-moons around GJ436b with different timing accuracies.

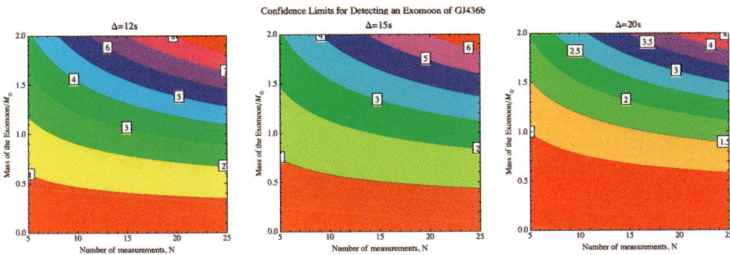

Figure 1. Confidence limits to which exomoons could be detected around the planet GJ 436b from N measurements. Here we show three possible timing accuracies given by Δ, where Δ represents the timing error on the mid-transit measurement. Each region represents the number of sigmas confidence to which the moon could be detected in integer steps.

5.1. An Earth-like Exomoon

One interesting hypothetical example to consider is an Earth-like exomoon around a Neptune-like planet. Consider a system like GJ436 but push the transiting planet back into a 47.3 day circular orbit. This would produce an equilibrium temperature of 273K. The Neptune-mass planet would be able to support an exomoon in a retrograde orbit up to a maximum of 0.9309 Hill radii according to the dynamical model of Domingos et al. (2006). Using the equations for a outward or inward migrating exomoon from Barnes & O'Brien (2002), a 1 $M_\oplus$ exomoon should be stable for at least 9.3 Gyr. If the exomoon was close to its maximal distance, it would generate a TTV of rms amplitude 11.4 minutes and TDV of rms amplitude 34 seconds using the expressions of Kipping (2009). Both of these signals should be quite detectable even with current instruments and thus allow for a determination of the exomoon mass and orbital separation. Furthermore, the probability this planet transiting is a respectable 1.2% and thus could easily be picked up in future surveys.

6. Conclusions

With characterisation of exoplanets taking a more prominent role in current and future astronomy, the search for exomoons will almost certainly step up as well. Some 400 years since the discovery of the Galilean moons, it is now possible to search for moons which are many light years away. It is impossible to predict what treasures of scientific discoveries will be made from such a search, but perhaps that is what makes the search so irresistible.

By employing transit time variation (TTV) and transit duration variation (TDV) in combination, it is possible to not only identify exomoons but also derive their mass and orbital separation from the host planet. Once the period, mass and ephemeris of an exomoon can be determined using these techniques, it should be possible to scrutinise the lightcurve and search for the dip in light due to the moon itself which would ultimately reveal the exomoon radius.

Acknowledgments. The author thanks STFC and UCL for supporting this research. Special thanks to Dr. Giovanna Tinetti without whom this work would not be possible.

References

Baines, E. K., van Belle, G. T., ten Brummelaar, et al. 2007, ApJ, 661, 577
Barnes, J. W. & O'Brien, D. P. 2002, ApJ, 575, 1087
Brown, T M., Charbonneau, D., Gilliland, R L., Noyes, R W., Burrows, A. 2001, ApJ, 552, 669
Cabrera, J. & Schneider, J. A&A, 464, 1133
Canup, R. M. & Ward, W. R. 2006, Nat, 441, 834
Deeg, H. J., 2002, ESA SP-514, 237
Domingos, R. C., Winter, O. C., Yokoyama, T. 2006, MNRAS, 373, 1227
Han, C. 2008, ApJ, 684, 684
Heyl, J. S. & Gladman, B. J. 2007, MNRAS, 377, 1511
Johnson, J. A., Winn, J. N., Cabrera, N. E., & Carter J. A. 2008, astro-ph/0812.0029
Jordan, A. & Bakos, G. A. 2008, accepted to ApJ, astro-ph/0806.0630
Kipping, D. M. 2008, MNRAS, 389, 1383
Kipping, D. M. 2009, MNRAS, 392, 181
Lathe, R., 2004, Icarus, 168, 18
Lewis, K. M., Sackett, P. D., Mardling, R. A. 2008, ApJ, 685, L153
Miralda-Escude, J. 2002, ApJ, 564, 1019
Ribas, I., Font-Ribera, A., & Beaulieu, J. P. 2008, ApJ, 677, L59
Pál, A. & Kocsis, B. 2008, MNRAS, 389, 191
Rafikov, R. R. 2008, submitted to ApJ, astro-ph/0807.0008
Reynolds, R. T., Squyres, S. W., Colburn, D. S., McKay, C. P. 1983, Icarus, 56, 246
Sartoretti, P. & Schneider, J., 1999, A&A, 14, 550
Scharf, C. A. 2006, ApJ, 648, 1196
Scharf, C. A. 2007, ApJ, 661, 1218
Seager, S., Kuchner, M., Hier-Majumder, C. A., Militzer, B. 2007, ApJ, 669, 1279
Simon, A., Szatmáry, K., Szabó, Gy. M., 2007, A&A, 470, 727S
Szabó, Gy. M., Szatmáry, K., Divéki, Zs. & Simon, A., 2006, A&A, 450, 395
Thommes, E. W., Matsumura, S., Rasio, F. A. 2008, Science, 321, 814
Tinetti, G. et al., 2007, Nature, 448, 169
Wolszczan, A. & Frail, D. A. 1992, Nature, 355, 145

Heather Knutson and Jo Harrington. Jo is wearing an exoplanet tie.

Ground Based Imaging Spectroscopy of Transiting Extrasolar Planets

Daniel Angerhausen and Alfred Krabbe

German SOFIA Institute, Institute of Space Systems, Pfaffenwaldring 31, 70569 Stuttgart, Germany

Abstract. We present results of an exploratory study to use near-infrared integral field spectroscopy to observe extrasolar planets. Our concept was tested with a K-Band time series observations of HD209458b and HD189733b obtained with SINFONI at the VLT and OSIRIS at Keck during secondary transits at a spectral resolution of R=3000. In this article we focus on the specific problems of ground-based, high signal-to-noise observations and demonstrate possible solutions for spetral timeseries. An advanced reduction method using elements of a spectral-differential decorrelation is discussed.

1. Introduction

Transiting exoplanets provide a unique opportunity for follow up exploration through phase-differential observations of their emission and transmission spectra. From such spectra immediate clues about the planets atmospheric composition and chemistry can be drawn. Such information is of imminent importance for the theory of the formation of planets in general as well as for their particular evolution.

Ground-based spectroscopy of exoplanet transits is a much needed extension of impressive results already obtained through space-based observations with Spitzer and HST.

The advantages of IFUs for ground-based NIR spectroscopy of Hot Jupiters were already discussed in prior publications (Angerhausen et al. 2006). Detailed analysis showed that the crucial challenge of the data reduction is to cope with the variable atmospheric conditions and systematic instrument effects during the observation (Angerhausen et al. 2007). Due to relatively short transit times the observation of telluric standards reduces the observing efficiency considerably, so that self-calibrating methods are preferred. Optimizing the observing efficiency with regard to the bright targets and short integration times on IFUs showed that observing in 'defocus' mode is the best choice.

2. Data Reduction

First step is the basic data reduction and extraction of the spectra. We set up a special calibration pipeline for such high S/N observations that is much more accurate than the standard pipelines provided for SINFONI and OSIRIS. Wavelength shifts and gross transmission differences were corrected for all spectra of the timeseries. After the standard reduction steps two different methods are used to analyse the broadband shape of

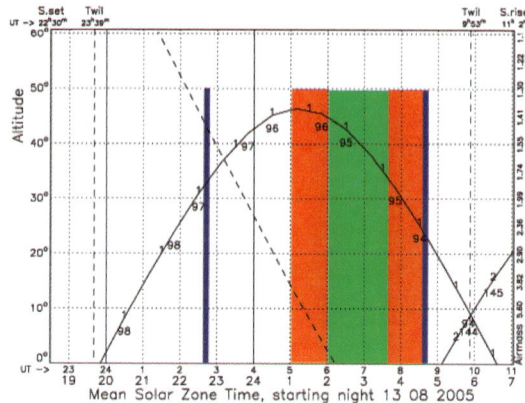

Figure 1. Schedule of an example observation with SINFONI at the VLT. Thick vertical (lue) lines represent start and end of our observation. The light shaded (green) area marks the phase of certain conjunction, dark shaded (red) areas represent phases of ingress (left) and egress (right). In and out of eclipse data can be observed at the same elevation/airmass.

the planetary spectrum (Fig. 2, left) and narrow band molecular spectral features (Fig. 2, right) respectively.

2.1. 'Direct' Method

For the analysis of the broadband spectrum pairs of in and out of eclipse spectra observed at the same airmass were divided. Therefore observation nights were chosen, that provided local zenith at either egress or ingress (Fig. 1). Averaging those spectra leads to a final broadband spectrum, that is independent of 1st order systematic effects due to airmass differences.

2.2. Decorrelation Method

For the analysis of narrow band features a spectral index $I_s = \frac{F_{\lambda_1} - F_{\lambda_2} + F_{\lambda_3}}{F_{\lambda_1} + F_{\lambda_2} + F_{\lambda_3}}$ is defined, with λ_2 position near an expected spectral feature, e.g. the 2.3 micron CO-feature predicted by almost all models (Fig 2, left). The timeseries of that index is decorrelated with any available observational parameters such as airmass, seeing, location of the PSF on the detector, humidity and local temperature (see Arribas et al. 2006, Fig. 3).

This method is a powerful tool to improve the standard deviation of the timeseries. It can theoretically correct any systematic that is correlated to a parameter, as long as that parameter is observed simultaneously. We are working on optimizing and automating the procedure and on a way to expand it so as to obtain a broadband spectrum by stepwise change of the central wavelength λ_2.

2.3. Modelling Atmospheric Transmission: ATRAN

ATRAN is a program that computes a synthetic/theoretical spectrum of atmospheric transmission (Lord 1992). It works for all infrared-wavelengths with selectable resolu-

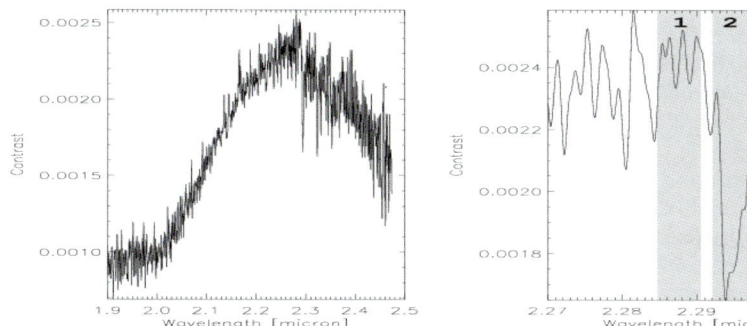

Figure 2. Expected HD189733b K-Band spectrum based on a modified spectrum by Barman (2005). Left: K-Band contrast spectrum for HD 189733b. For analysis of the broad band spectrum pairs of in and out of eclipse spectra taken at the same airmass were compared. Right: Narrow band CO-feature at 2.29 micron, grey areas and numbers highlight the wavelength ranges used to define the spectral index $I_s = (F_1 - F_2 + F_3)/(F_1 + F_2 + F_3)$. A timeseries of this Index was analysed in a spectral differential and decorrelation method.

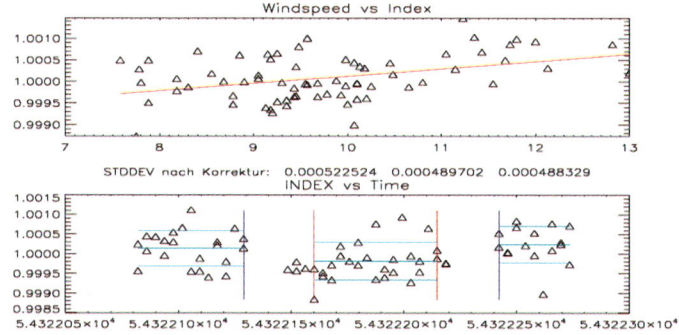

Figure 3. Example of one decorralation step. Top: The spectral index I_s is plotted against the observational parameter windspeed and decorrelated by a linear fit. Bottom: time-series of spectral index with improved standard deviation (light horizontal blue bars) after decorrelation (between the middle pair of vertical red lines: in eclipse, left and right: out of eclipse).

tion. Input parameters are location of the observatory, zenith angle and the atmospheric concentration of various trace gases (methane, CO, CO_2, water etc).

From our experience ATRAN is an essential tool for ground-based observations. It can be used to compute wavelength-dependent transmission filters and correlations between spectra and particular telluric species.

Our idea for an improved data reduction concept is to use ATRAN to fit an atmospheric model to each individual observation.

Table 1. Pearson correlation coefficient for a set of observational parameters with the spectral index before and after airmass decorrelation

Observational parameter	Coefficient before airmass decorr.	Coefficient after airmass decorr.
Airmass	0.994045	0.000775055
Windspeed	-0.567896	0.433901
Wavelength-shift	-0.348878	0.383352
Humidity	-0.638413	0.370100
Temperature	0.652521	-0.345674
Winddirection	-0.0936889	0.112939
Strehl Ratio	0.588461	-0.00937599
Seeing	0.474509	-0.00885338
Pressure	-0.909824	0.131780
X-center PSF	-0.515934	0.107597
Y-center PSF	0.631093	-0.320321
FWHM in X-dir.	0.211757	-0.147461
FWHM in Y-dir. FIT	0.192949	-0.187359

Another approach is to use ATRAN models to compute timeseries of narrow telluric line features to use them as decorrelation parameter (see section 2.2) for estimating and removing variable absorption by atmospheric trace molecules.

3. Results and Perspectives

Our study showed that we can reach the necessary levels of signal to noise of about 10^4 to detect the planetary signal. Instrument systematics, changing airmass and even noise induced by weather parameters can be decorrelated with the described method. Nevertheless the central problem of variability in concentration of atmospheric trace gases, due to changing diurnal photochemistry has to be solved. We are currently working on the above mentioned methods to also decorrelate those effects.

References

Angerhausen D., Krabbe, A., & Iserlohe C. 2006, in: Proc. SPIE, Ground-based and Airborne Instrumentation for Astronomy, 6269, ed. McLean, I.S., & Iye, M., 4
Angerhausen D., Krabbe A. & Iserlohe C. 2007, in: ASP Conf. Ser., Vol. 366 Transiting Extrasolar Planets Workshop, ed. C. Afonso, D. Weldrake, & Th. Henning, 262
Arribas, S. et al. 2006, PASP, 118, 839, 21-36
Barman, T. 2005, ApJ, 632, 1132-1139
Lord S. D. 1992, NASA Technical Memorandum 103957

Ground-based Detection of the Secondary Eclipse of TrES-3b

E.J.W. de Mooij and I.A.G. Snellen

Leiden Observatory, P.O. Box 9513, NL-2300 RA Leiden, The Netherlands

Abstract. Secondary eclipse measurements provide information on the thermal properties of exoplanet atmospheres. Here we summarise the results of ground-based K-band observations of the secondary eclipse of the planet TrES-3b (see de Mooij & Snellen 2009). The observations were made using the LIRIS instrument on the William Herschell Telescope. The secondary eclipse was detected at a level of -0.241±0.043% (~6σ), corresponding to a brightness temperature of 2040±125 K.

1. Introduction

Secondary eclipse measurements of transiting extrasolar planets with the Spitzer Space Telescope have yielded several direct detections of thermal exoplanet light. However, arguably one of the most interesting parts of the planet spectrum (1-3μm) is inaccessible with this satellite. This wavelength region is at the peak of the planet's spectral energy distribution and is also the regime where molecular absorption bands can significantly influence the measured emission. So far, only the recent observations using the HST (Swain et al. 2009) have let to detections in this wavelength range. We summarise the results of the first ground-based detection of the secondary eclipse of the very hot Jupiter TrES-3b (O'Donovan et al. 2007) in K-band, using the LIRIS instrument on the William Herschell Telescope (WHT).

2. Observations

We have used the William Herschel Telescope to observe the secondary eclipse of TrES-3b in July 2008. The observations were made in K-band using the LIRIS infrared instrument and lasted the entire night. A total of 1800 frames were obtained, of which ~330 were taken during the eclipse. To increase the efficiency and reduce intrapixel variations, we significantly defocused the telescope in such a way that the PSF took the shape of an annulus with a radius of approximately 4 pixels.

3. Data Reduction

Here we give a short summary of the data reduction (the complete data reduction process is described in de Mooij & Snellen 2009). First the images were corrected for crosstalk between the different quadrants of the detector and for its non-linear behaviour. After these corrections the data were flatfielded using a flatfield created from a series of domeflats. Subsequently, aperture photometry was performed on TrES-3b

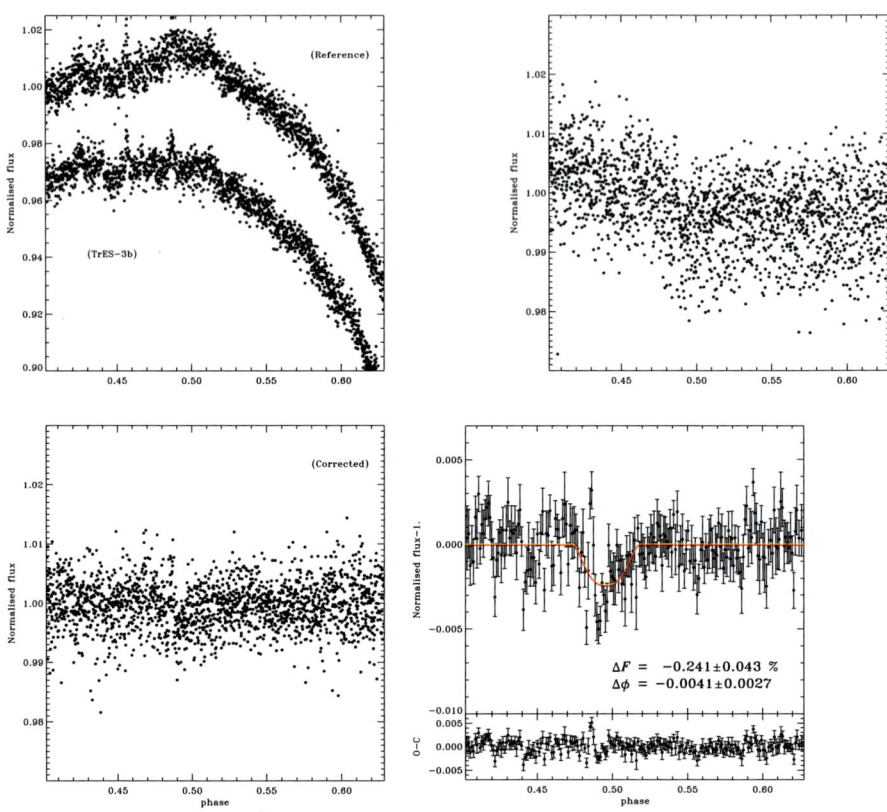

Figure 1. Top left: The raw lightcurves of TrES-3b and a nearby reference star. Top right: The lightcurve of TrES-3b normalised using the lightcurve of the reference star. Bottom left: the normalised raw lightcurves of TrES-3b and the reference star. Bottom right: the 9-point binned lightcurve of the secondary eclipse of TrES-3b. The overplotted line is the best fitting model.

and a K=9.77 reference star, using an aperture of R=13 pixels (~3.25"). The resulting lightcurves are shown in the top left panel of figure 1.

4. Analysis and Results

After the aperture photometry, we divided the lightcurve of TrES-3b by that of the reference star, the result is shown in the top right panel of figure 1. The measured flux is correlated with the stars' (x,y)-position on the detector. This correlation is the source of the gradient seen in normalised lightcurve in the top right panel of figure 1. We corrected for this effect by fitting a linear function to the correlation outside of the eclipse, and applying this to the entire lightcurve. We show this corrected lightcurve in the bottom left panel of figure 1. The bottom right panel of this figure shows the same, corrected, lightcurve binned by 9 points (1 dither cycle).

We measure a rms-noise out of eclipse of 0.0038, which is ~2 times the photon noise of 0.0015. To measure the eclipse depth, we fitted an analytical model by Mandel & Agol (2002), using parameters for the TrES-3b system from Sozzetti et al. (2008). We measure the secondary eclipse depth of TrES-3b to be $-0.241\pm0.043\%$ (~6σ), where correlated noise is taken into account in the uncertainties. This eclipse depth corresponds to a day-side brightness temperature of $T_b(2.2\mu m)=2040\pm125$ K, which is consistent with current models by Forney et al. (2008) of this planet's upper atmosphere. The centre of the eclipse seems to be slightly offset from phase $\phi=0.5$ by $\Delta\phi=-0.0041\pm0.0027$, which could indicate that the orbit of TrES-3b is non-circular.

5. Future possibilities

These observations, together with those of Sing & Lopez-Morales (2009), show that it is possible to use ground-based telescopes to detect the thermal emission from hot Jupiters. It opens up the possibility to investigate the near-infrared and optical spectral energy distribution of exoplanets. Not only does the near-infrared region contain a very exciting part of the SED due to the influence of molecular bands ground-based observations in the near infrared will also be able to complement the warm mission of the Spitzer Space Telescope by expanding the wavelength coverage.

Acknowledgments. Based on observations made with the William Herschell Telescope operated on the island of La Palma by the Isaac Newton Group in the Spanish Observatorio del Roque de los Muchachos of the Instituto de Astrofísica de Canarias.

References

Fortney et al. 2008, ApJ 678, 1419
Mandel & Agol 2002, ApJ 680, L171
de Mooij, E.J.W. & Snellen, I.A.G. 2009 A&A, 493, L31
O'Donovan et al. 2007, ApJ 663, L37
Sing, D.K. & Lopez-Morales, M. 2009 A&A, 493, L35
Sozzetti et al. 2008, arXiv:astro-ph/0809.4589
Swain et al. 2009, ApJ 690, L114

Yuk Yung.

Exoplanet Spectroscopy: The Hubble Case

Pieter Deroo,[1] Mark Swain,[1] Gautam Vasisht,[1] Pin Chen,[1] Giovanna Tinetti,[2] Jeroen Bouwman,[3] Daniel Angerhausen,[4] and Yuk Yung[5]

Abstract. The Hubble Space Telescope has recently emerged as the first telescope to detect molecular signatures in an exoplanet via infrared spectroscopy. Molecular spectroscopy of exoplanets is demanding and requires an accurate determination and removal of the instrument systematics. Here we report on our effort to extract accurate exoplanet spectra from NICMOS spectrophotometry. We developed a standardized and highly automated pipeline to remove instrument systematics based on our previous results. We tested the pipeline and find excellent agreement with observation specific implementations. The process of decorrelating instrument parameters from the measured time series is well understood, stable and guarantees reproducible results.

1. Introduction

The first spectroscopic detection of molecular signatures in the infrared transmission spectrum of HD 189733b (Swain et al. 2008) was a dramatic demonstration of the possibility of exoplanet spectroscopy with sufficient precision to identify the presence and quantify the abundance of oxygen and carbon bearing molecules. Now, also the dayside emission spectrum is measured (Swain et al. 2009) and water (H_2O), methane (CH_4), carbon monoxide (CO), and carbon dioxide (CO_2) are identified through spectroscopy in HD 189733b.

The strategy for extracting the spectrum of an eclipsing exoplanet consists of measuring a spectrophotometric time series pre, during and post eclipse. By comparing the flux measured in eclipse with that obtained out of eclipse, we construct a planetary spectrum. Although HST avoids limitations imposed by the Earth's atmosphere, spectrophotometry with NICMOS is subject to systematic errors that must be corrected because they are comparable in size to the expected molecular signatures. A complete description of the data-analysis strategy is described in the supplementary information of Swain et al. (2008) and updates are provided in Swain et al. (2009).

In this proceeding, we present the results of our effort to extract high accuracy near-IR spectra of transiting exoplanets with the NICMOS camera. The analysis is

---

[1]Jet Propulsion Laboratory, 4800 Oak Grove Drive, Pasadena, CA 91109

[2]University College London, Gower Street, London WC1E6BT, UK

[3]Max-Planck Institute for Astronomy, Konigstuhl 17, D-69117 Heidelberg, Germany

[4]German SOFIA institute at the Institute of Space Systems, Universität Stuttgart, Pfaffenwaldring 31, 70569 Stuttgart

[5]Division of Geological and Planetary Sciences, California Institute of Technology, Pasadena, CA 91125

64   Deroo et al.

based on our previous results but the reduction is now treated as a pipeline, such that the reduction is standardized and automated.

2. Reduction Pipeline Description

The reduction starts from the time series of image files processed by the STScI pipeline to include all NICMOS standard reductions except flat-fielding (i.e. the CAL extension output). To increase the saturation time of an individual pixel and provide better stability (i.e. averaging over intra-pixel and pixel quantum efficiency is possible), the camera is typically defocused.

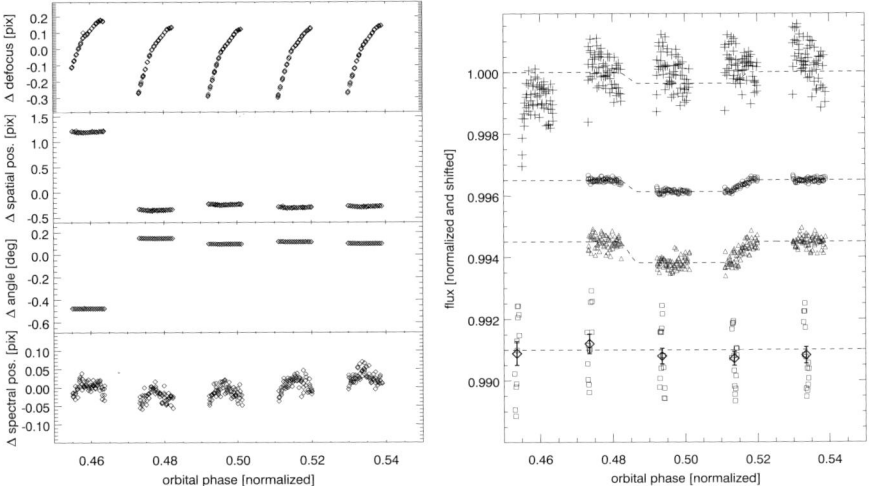

Figure 1.   left: The time series of the optical state vectors for a typical observation (here shown for HD 209458). right: Normalized light curves; from top to bottom: (i) the raw broadband light curve (ii) the calibrated broadband light curve (iii) the calibrated 1.9 μm light curve (iv) contemporaneous MOST visible photometry data, showing both individual and averaged data.

As a first step, we generate an observation-specific flat-field taking into account the color of the illuminated pixels. We use a set of recent on-orbit flats to fit the wavelength dependence for each pixel as a quadratic function. We establish the wavelength for each column of the observation from the calibration exposures and extract the resulting quadratic at the appropriate wavelength by column and apply this correction to the observations. We note that a flat-field correction might be considered optional for differential photometry, since we are only interested in differential changes in time. There are, however, small offsets between different exposures (see further), for which good flat fielding should reduce sensitivity to pixel-to-pixel quantum efficiency changes.

Next, we correct the few bad pixels in the spectrum and do a background correction by fitting unilluminated regions adjacent to the spectrum. Then we determine the *state of the observation*. We fit Gaussians along the spatial direction for each column to determine the width and mean position of the spectrum. These multi-column fits are used to quantify, for each exposure, the angle and position of the spectrum on the detector as well as a proxy to the defocus. The data is now *homogenized*: we derotate and shift the spectrum to place all observations in a common, pixel-based reference frame. A spatial mask is applied and spectra are extracted by summing in the (new) spatial direction. When multiple orders are present on the detector, the orders are co-added to improve signal-to-noise. As a last step in the extraction method, we determine the displacement in the spectral direction by identifying differential shifts from the edges of the transmission curve.

Time series of the optical state vectors (defocus, position, angle, spectral shift) are shown in Fig. 1. Our assumption is that the light curves have causal connection with these state vectors. Furthermore, we assume that the light curve behavior can be described by perturbations which are linear in these variables. It is clear from Fig. 1, that the position and angle of the spectrum on the detector in the first orbit differs strongly from the other orbits. We, therefore, exclude the first orbit data as a standard precaution in any further analysis, since the linear approach could work poorly for these data.

Following Swain et al. (2008), we correct for instrumental systematics based on the behavior of the optical state vectors, the orbital phase and its square and the temperature of the detector. We do this by using a downhill-simplex method to minimize the residuals in the light curve. We tested the difference with a Gauss-Markov method (as in Swain et al. 2008) and find that both methods provide similar results, but a downhill-simplex method is easier to customize and faster. The standard approach of the pipeline is to do a joint decorrelation of the in and out of eclipse data, although it is also capable of the approach used in Swain et al. (2008), where a model for the instrument systematics is derived from the out-of-eclipse data and interpolated to the in-eclipse section. In the joint decorrelation approach, we provide orbital parameters to the code and it will generate results for the free parameters. The approach is to compute the corrected light curve for trial decorrelations of the instrument systematics and trial model light curves. Minimizing the power in the residuals yields the desired parameters (e.g. planetary radius). We note that our implementation of the decorrelation algorithm allows us to increase the number of state vectors and to do a multi-parameter light curve modeling. Our strategy is, however, always to use the minimum number of variables possible, not to include non-causal parameters or to do over fitting. An example of the decorrelation process is shown in Fig. 1, where normalized light curves before and after correction are shown.

3. Verification of the Pipeline

The data-analysis approach presented here remains largely unchanged from Swain et al. (2008, 2009), but is updated in three significant manners. First the extraction of the spectrum from the time-series has become an automated process. Second, the decorrelation of the instrument parameters is now fully pixel based rather than average pixel based, as in the past. The advantage is that pixel sensitivities are better sampled and that we can work with the full, although oversampled, spectral resolution of the instrument. Third, we now derotate and shift the spectrum to put the observations in a single reference frame.

We tested the new reduction and decorrelation package by re-computing the transmission and dayside emission spectrum of HD 189733b and compare it to the published results in Swain et al. (2008, 2009), which is shown in Fig. 2. Both methods provide

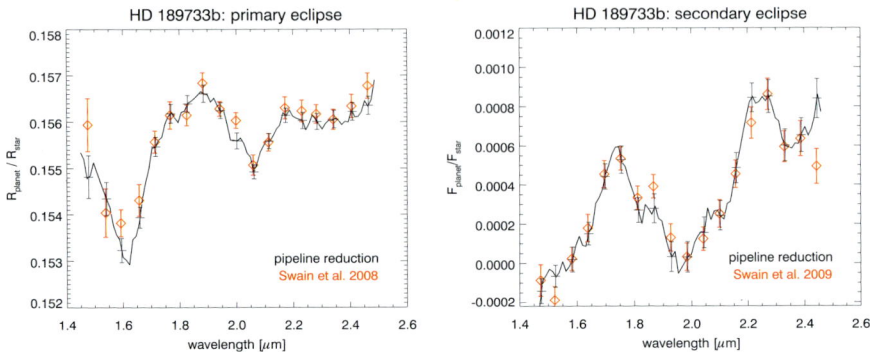

Figure 2. The code was verified by re-computing the transmission (*left*) and dayside emission spectrum (*right*) of HD 189733b.

excellent agreement, with differences between the spectra within the statistical uncertainty of the measurements.

4. Conclusions

Measuring a high signal-to-noise exoplanet spectrum with NICMOS is demanding and requires a proper modeling and removal of the systematics involved. The detection of molecular signatures in the transmission and dayside emission spectra of HD 189733b are a direct confirmation of our ability to achieve the required stability. With the development of our reduction pipeline, we demonstrate that the data analysis method, based on decorrelation of instrument parameters from the measured spectrophotometric time series, is well understood and stable. Our analysis approach remained largely unchanged except for it becoming highly automated, pixel based and placing all observations in a common-pixel based reference frame. Moreover, the decorrelation algorithm has been reimplemented to become more extendable. We tested the pipeline by re-reducing the observations of HD 189733b and find excellent agreement.

Acknowledgments. Pieter Deroo is grateful for NASA Fellowship.

References

Swain, M. R., Vasisht, G. V., & Tinetti, G. 2008, Nat, 452, 329
Swain M. R. et al. 2008, ApJ, 960, L114.

David Charbonneau.

Virginie Batista and the tools of a chairman.

Part III

Exoplanet Atmospheric Dynamics (Chairs: Caroline Terquem and Adam Showman)

Atmospheric Dynamics of Two Eccentric Transiting Planets: GJ 436b and HD 17156b

N. K. Lewis and A. P. Showman

Department of Planetary Sciences and Lunar and Planetary Laboratory, The University of Arizona 1629 University Blvd., Tucson, AZ 85721 USA

J. J. Fortney

Department of Astronomy & Astrophysics, UCO/Lick Observatory, University of California, Santa Cruz, CA 95064 USA

M. S. Marley and R. S. Freedman

NASA Ames Research Center 245-3, Moffett Field, CA 94035 USA

Abstract. Extrasolar planets on eccentric orbits present a unique opportunity to study the effects of variable heating and non-synchronous rotation on the atmospheric dynamics of hot Jupiters and hot Neptunes. We present three-dimensional atmospheric circulation models that include realistic radiative transfer for two such extrasolar planets: GJ 436b (e=0.15) and HD17156b (e=0.67). GJ436b is one of the smallest transiting extrasolar planet known to date. Because of its size, it is likely to have an atmospheric composition more similar to Neptune (~30x Solar), which has an effect on the radiative transfer and hence dynamics of the planet's atmosphere. HD17156b is a fairly massive Jupiter sized planet on a highly elliptical orbit. During its orbit, HD17156b passes through the radiative regime of both pM and pL Class planets as defined by Fortney et al. (2008), which makes it an ideal candidate to test the effects of TiO and VO in the atmospheres of extrasolar planets. We contrast the global circulation patterns and vertical thermal structures of these planets to those of HD189733b and HD209458b and postulate possible observational implications.

1. Introduction

To date, 335 extrasolar planets have been detected, ~78 of which have been determined to have significantly non-zero eccentricities (Schneider 2009). This population of eccentric extrasolar planets is interesting for several reasons. Tidal evolution theory predicts that since many of these planets are very close to their host stars (within ~0.2 AU) their orbits should circularize over time. This can help to constrain tidal Q parameters and ages for these planetary systems as well as test current orbital dynamics models and their ability to explain extrasolar planetary orbital configurations (Jackson et al. 2008). Additionally, eccentric extrasolar planets inhabit an interesting thermal regime where the time variability in the distance of a given planet from its host star will

---
[1]SETI Institute, 515 North Whisman Road, Mountain View, CA 94043 USA

result in variable stellar heating. Perhaps the most interesting of the eccentric extrasolar planets are those that transit their host star as seen from Earth. Observations of transiting extrasolar planets can provide constraints on planetary atmospheric structure, composition, and thermal emission. These observational constraints can then be used in atmospheric models to better understand the possible atmospheric dynamics of these planets in interesting thermal forcing regimes.

Here, we present numerical simulations using a coupled radiative transfer and three-dimensional general circulation model (GCM) to model the atmospheres of two transiting eccentric extrasolar planets: GJ436b and HD17156b. The following sections review the details of the atmospheric model employed in this study and present the relevant planetary and stellar parameters for the GJ436 and HD17156 systems. Additionally, results from the atmospheric simulations are presented along with a discussion of the significance of these findings and suggestions for further work.

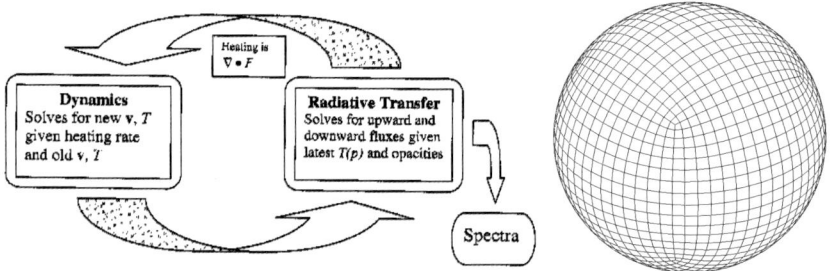

Figure 1. Left Panel: Flow diagram outlining the interrelationship between the radiative transfer and general circulation model. Right Panel: Cubed-sphere grid at 16x16 resolution on each cubed face (C16).

2. Model

The atmospheric model used in this study is a coupled radiative transfer and dynamics model that was specifically developed with the study of extrasolar planetary atmospheres in mind. The Substellar and Planetary Atmospheric Radiation and Circulation (SPARC) model is described in detail in Showman et al. (2009). The SPARC model employs the MITgcm (Adcroft et al. 2004) to treat the atmospheric dynamics and the model of Marley & McKay (1999) to treat the radiative transfer. The coupling of these two models is presented graphically in Figure 1. The MITgcm solves the 3D primitive equations in a spherical geometry, which in this study is partitioned using the cubed sphere grid (Fig. 1). Between 40 and 53 atmospheric layers are used for the vertical gridding of the atmosphere spanning pressures, p, between 200 bar to as low as 2 μbar with even $\log(p)$ spacing of the layers. A two-stream variant of the plane parallel radiation transfer scheme of Marley & McKay (1999) is implemented to calculate the upward and downward radiative fluxes. The chemical abundances and wavelength-dependent opacities are pre-tabulated over a large grid in pressure and temperature, as described in Freedman et al. (2008). In the simulations presented here, the dynamics are updated every 10-50 seconds while the radiative transfer is only updated every 200 seconds for computational efficiency.

3. Planetary Parameters

This study focuses on two eccentric transiting extrasolar planets: GJ436b and HD17156b. Planetary and stellar parameters for these two systems are outlined in Tables 1 and 2 respectively. GJ436b and HD17156b represent extremes in size and eccentricity, respectively, for the transiting extrasolar planet population. Because both of these planets are on eccentric orbits, the effects of non-synchronous rotation and time-varying distance from the host star were incorporated into the SPARC model (Showman et al. 2009). The most probable rotation rate of the planet has been determined using the following relationship presented in Hut (1981):

$$P_{rot} = P_{orb} \left[\frac{(1 + 3e^2 + \frac{3}{8}e^4)(1 - e^2)^{3/2}}{1 + \frac{15}{2}e^2 + \frac{45}{8}e^4 + \frac{5}{16}e^6} \right] \quad (1)$$

where P_{rot} is the planetary rotation rate, P_{orb} is the orbital period of the planet, and e is the eccentricity of the planetary orbit. In all cases cases considered here the obliquity of the planet is assumed to be zero. The time-varying distance of the planet with respect to its host star, $r(t)$, can be determined using Kepler's equation (Murray & Dermott 1999).

Table 1. Planetary Data

Planet	M_P [M_J]	R_p [R_J]	a [AU]	e	g [m/s^2]	P_{orb} [days]	P_{rot} [days]
GJ436b[1]	0.071	0.437	0.028	0.15	9.22	2.643904	2.328553
HD17156b[2]	3.09	1.23	0.1589	0.6719	50.6	21.21747	3.76797

REFERENCES: (1) Bean et al. (2008); (2) Gillon et al. (2008)

Table 2. Stellar Data

Star	Spectral Type	[Fe/H]	T_{eff} [K]	M_* [$M_\odot$]	R_* [$R_\odot$]	log g [cgs]
GJ436a[1,2]	M2.5	-0.03	3200	0.44	0.505	4.85
HD17156a[3,4]	G0 V	0.24	6079	1.2	1.63	4.09

REFERENCES: (1) Maness et al. (2007); (2) Bean et al. (2008); (3) Fischer et al. (2007); (4) Gillon et al. (2008)

4. Results: GJ436b

Because GJ436b is similar in size to Neptune, it is also likely to have a similar composition. The inferred heavy element composition of Neptune's atmosphere from methane measurements is between 30x and 60x solar values (Gautier et al. 1995). Given these constraints for Neptune, simulations of the atmospheric dynamics of GJ436b were run for both 1x and 30x solar composition atmospheres. The top panels of Figure 2 show a comparison of the temperature and winds patterns at the 60 mbar level between the 1x solar and 30x solar composition simulations. The 8 μm photosphere in the 1x solar

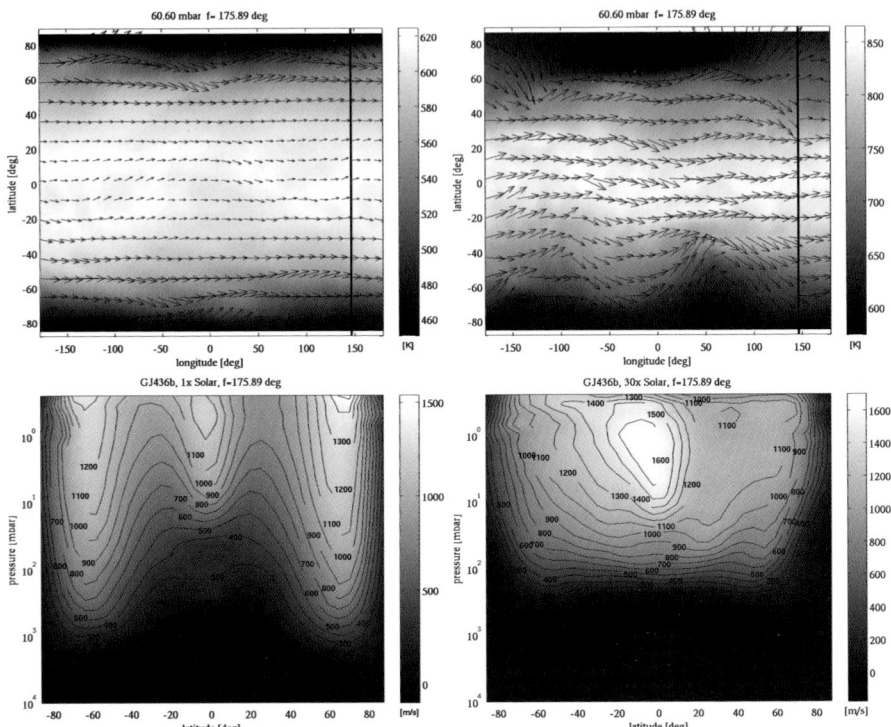

Figure 2. *Top*: Temperature (greyscale) and winds (arrows) at 60 mbar for the 1x (*left*) and 30x (*right*) solar composition GJ436b simulations. The longitude of the substellar point in each panel is indicated by the solid vertical line. *Bottom*: Zonal-mean zonal winds (greyscale) for the 1x (*left*) and 30x (*right*) solar composition GJ436b simulations. Resolution is C32 (approximately 128x64 in longitude and latitude) with 40 vertical layers covering a range of pressures from 200 bar to 0.2 mbar.

case should be near the 60 mbar level as was determined for HD189733b in Showman et al. (2009). There are several distinct differences between the temperatures and wind patterns in the 1x solar and 30x solar composition cases for GJ436b. First, the 30x solar case is on average around 200 K warmer than the 1x solar case at the 60 mbar level. Second, the equatorial winds in the 30x solar case are stronger relative to the high-latitude winds when compared with the 1x solar case. Lastly, the equator-to-pole temperature difference is around 100 K greater in the 30x solar case than in the 1x solar case. Neither the 1x solar nor the 30x solar GJ436b simulations show any clear "hot spots" near the level of the photosphere at any point in the orbit. However, small ($\sim 10-20$ K) temperature variations can be seen over the course of the planetary orbit as well as from orbit to orbit. These temperature variations are likely to be linked to the pressure and temperature dependent atmospheric radiative timescales expected for GJ436b (Fortney et al. 2008).

Increases in the latitudinal temperature gradient correspond to increases in the vertical gradient of the east-west wind speeds according to the thermal wind equation

(Holton 2004). The bottom panels of Figure 2 show the vertical profile of the longitudinally averaged east-west winds, or zonal-mean zonal winds, for both the 1x solar and 30x solar GJ436b simulations. Overall, the vertical gradient of the zonal winds is larger for the 30x solar composition case than for the 1x solar composition case, which follows from the equator-to-pole temperature differences seen in Figure 2. Additionally, the maximum wind speeds in the upper layers of the simulation in the 30x solar composition case exceed those of the 1x solar case. The overall shape of the jet structures that developed in the 1x solar and 30x solar simulations are also quite different with the 1x solar case having a three jet structure with strong high-latitude winds and the 30x solar case having a single strong equatorial jet. Further investigation is needed to determine the mechanisms that might cause differences between the zonal winds for the 1x solar and 30x solar abundance cases of GJ436b.

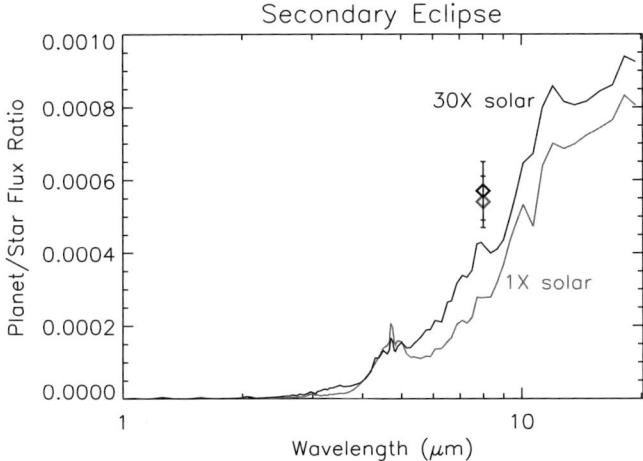

Figure 3. Planet/Star flux ratio for secondary eclipse calculated for both the 1x and 30x solar composition cases of GJ436b. The 8 μm secondary eclipse observations of Deming et al. (2007) and Demory et al. (2007) are indicated by the black and grey diamonds respectively.

For GJ436b, additional calculations were performed to determine the emergent flux of the planet as a function of wavelength at various orbital phases according to the methodology presented in Fortney et al. (2006). Figure 3 shows the predicted planet/star flux ratio in the 1-20 μm wavelength range for the 1x solar and 30x solar cases of GJ436b during secondary eclipse along with current 8 μm measurements. Although the 30x solar case does show an increase in the planet/star flux ratio over the 1x solar case, it still does not explain current secondary eclipse measurements for GJ436b. It is likely that other factors such as tidal heating or disequilibrium chemistry must be included in atmospheric models in order to explain current observations.

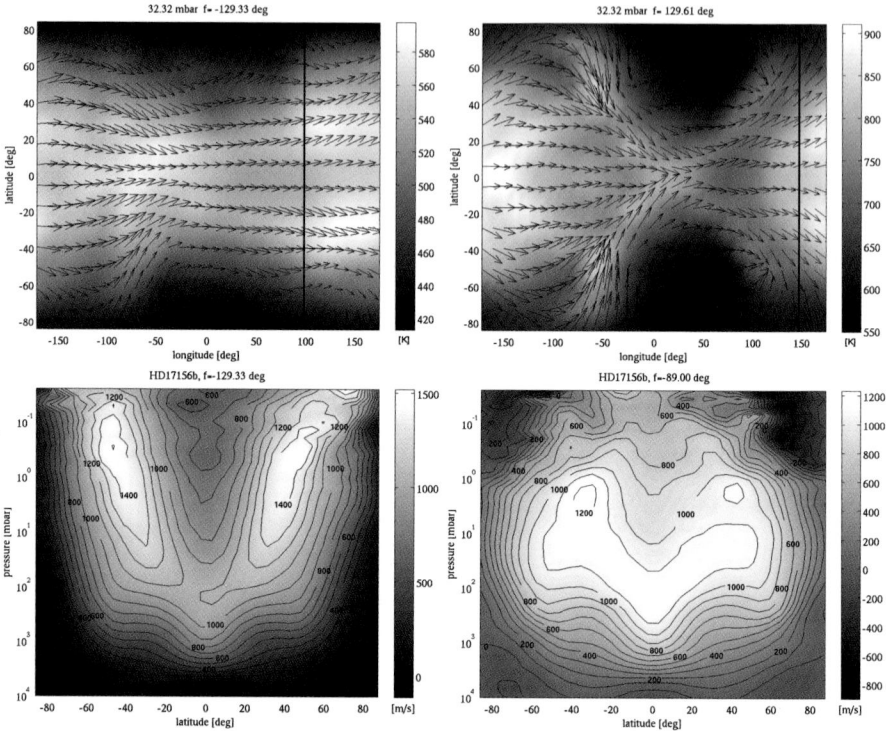

Figure 4. *Top*: Temperature (greyscale) and winds (arrows) for the 32 mbar level of the 1x solar composition with TiO and VO HD17156b simulation. The left panel corresponds to a planetary true anomaly, f, of 129.33° before periapse passage ($f = 0°$). The right panel corresponds to $f = 129.61°$ after periapse passage. The longitude of the substellar point in each panel is indicated by the solid vertical line. *Bottom*: Zonal-mean zonal winds (greyscale) at two points in the orbit of HD17156b. The left panel corresponds to well before periapse at $f = -129.33°$. The right panel corresponds to the zonal wind pattern that develops as the planet begins to approach periapse ($f = -89.00°$). Note that the upper and lower right panels do not correspond to the same point in the orbit of GJ436b, but simply illustrate extremes in thermal and wind structure variations seen in the simulations. Resolution is C16 (approximately 64x32 in longitude and latitude) with 47 vertical layers covering a range of pressures from 200 bar to 20 μbar.

5. Results: HD17156b

HD17156b is a fairly massive ($\sim 3\ M_J$) Jupiter sized planet on a highly eccentric ($e = 0.6719$) orbit. At periapse, HD17156b is within 0.052 AU of its host star and receives enough incident flux to be considered a 'pM Class' planet with the possibility of a TiO/VO induced stratosphere as postulated in Fortney et al. (2008). The large swings in stellar insolation received by HD17156b during its orbit results in interesting periodic changes in the thermal and wind profiles of the planet as shown in Figure 4. The top panels of Figure 4 show the thermal and wind profile of HD17156b at the 30 mbar level before and after periapse passage. After periapse passage, the planet shows a significant (~300 K) increase in overall temperature. Interestingly, the points in the orbit where the planet is warmest and coolest at the 30 mbar level do not occur directly at periapse and apoapose, but some time (~ 3 days) after closest and furthest approach from its host star. From the bottom panels of Figure 4 it is also clear that the zonal jet structure and maximum wind speeds change along the orbit of HD17156b. These predicted strong changes in temperature and winds as HD17156b passes into and out of periapse could significantly impact measured light curves and spectra. Although HD17156b does transit its host star, the probability of observing a secondary eclipse for this system is close to zero (Gillon et al. 2008). HAT-P-2b with its short orbital period (~ 5.6 days), large eccentricity ($e = 0.50$), and the possibility for full orbit observations may prove to be a better test for changes in the wind and temperature structure of eccentric planets near periapse.

6. Discussion

This study has taken a first look at the possible atmospheric dynamics and thermal structures for eccentric extrasolar planets using a coupled three-dimensional dynamics and radiative transfer model. It is clear from these simulations of GJ436b and HD17156b that eccentric extrasolar planets may have different wind and thermal profiles than what have been predicted for non-eccentric planet such as HD189733b and HD209458b. The non-synchronous rotation rates expected for eccentric planets favors the development of mid-to-high latitude jets, which is in contrast to the strong equatorial jet that develops in synchronously rotating cases (Showman et al. 2009). The GJ436b simulations presented here tend to have longitudinally homogenized temperature distributions near the level of the photosphere. This temperature homogenization means that there may not be a clear hot and cold spot that would be detectable in a full orbit light curve, but instead the light curve might reflect the gradual increase and decrease of the overall planetary temperature profile as the distance of the planet from its host star changes with time. The HD17156b simulations show strong temporal variation in both the thermal and wind structures in the atmosphere. Thermal hot/cold spots may be detectable in a partial light curve for HD17156b, but the strength and location of the hot/cold spots could change throughout the planet's orbit and even from orbit to orbit.

Eccentric extrasolar planets offer the opportunity to study atmospheres that are not expected to have steady-state winds and temperature profiles. Continued iterations of model atmospheres to help explain and perhaps predict extrasolar planet observations will enhance not only the current understanding of extrasolar planet atmospheres, but also atmospheric dynamics and radiative transfer in general. Future work with the SPARC model includes exploring the effects of tidal heating in eccentric extrasolar planet atmospheres and further application of current models to other extrasolar planets with *Spitzer* multi-wavelength data. Although close-up images of these giant planets orbiting far away stars may be in the distant future, continued atmospheric modeling efforts will help to paint a clearer picture of these worlds unlike anything in our own solar system.

Acknowledgments. This study was supported by NASA Origins grant NNX08AF27G and NASA NESSF grant NNX08AX02H.

References

Adcroft, A., Campin, J.-M., Hill, C., & Marshall, J. 2004, Monthly Weather Review, 132, 2845
Bean, J. L., Benedict, G. F., Charbonneau, D., Homeier, D., Taylor, D. C., McArthur, B., Aeifahrt, A., Dreizler, S., & Reiners, A. 2008, A&A, 486, 1039
Deming, D., Harrington, J., Laughlin, G., Seager, S., Navarro, S. B., Bowman, W. C., & Horning, K. 2007, ApJ, 667, L199
Demory, B.-O., Gillon, M., Barman, T., Bonfils, X., Mayor, M., Mazeh, T., Queloz, D., Udry, S., Bouchy, F., Delfosse, X., Forveille, T., Mallmann, F., Pepe, F., & Perrier, C. 2007, A&A, 475, 1125
Fischer, D. A., Vogt, S. S., Marcy, G. W., Butler, R. P., Sato, B., Henry, G. W., Robinson, S., Laughlin, G., Ida, S., Toyota, E., Omiya, M., Driscoll, P., Takeda, G., Wright, J. T., & Johnson, J. A. 2007, ApJ, 669, 1336
Fortney, J. J., Cooper, C. S., Showman, A. P., Marley, M. S., & Freedman, R. S. 2006, ApJ, 652, 746
Fortney, J. J., Lodders, K., Marley, M. S., & Freedman, R. S. 2008, ApJ, 678, 1419
Freedman, R. S., Marley, M. S., & Lodders, K. 2008, ApJS, 174, 504
Gautier, D., Conrath, B. J., Owen, T., de Pater, I., & Atreya, S. K. 1995, in Neptune and Triton (Tucson: The University of Arizona Press), 547
Gillon, M., Demory, B.-O., Barman, T., Bonfils, X., & Queloz, D. 2007, A&A, 471, L51
Gillon, M., Triaud, A. H. M. J., Mayor, M., Queloz, D., Udry, S., & North, P. 2008, A&A, 485, 871
Holton, James R. 2004, An Introduction to Dynamic Meteorology (Burlington, MA: Elsevier)
Hut, P. 1981, A&A, 99, 126
Jackson, B., Greenberg, R., & Barnes, R. 2008, ApJ, 678, 1396
Maness, H. L., Marcy, G. W., Ford, E. B., Hauschildt, P. H., Shreve, A. T., Basri, G. B., Butler, R. P., & Vogt, S. S. 2007, PASP, 119, 90
Marley, M. S., & McKay, C. P. 1999, Icarus, 138, 268
Murray, C. D., & Dermott, S. F. 1999, Solar System Dynamics (New York, NY: Cambridge University Press)
Schnider, J. 2009, The Extrasolar Planets Encyclopaedia, http://exoplanet.eu
Showman, A. P., Fortney, J. J., Lian, Y., Marley, M. S., Freedman, R. S., Knutson, H. A., & Charbonneau, D. 2009, preprint (arXiv:0809.2089v1)

Giovanna Tinetti and Ofer Lahav sharing one of the mystery wines at the reception.

Part IV

Molecular Data-lists and Modelling (Chair: Bruno Bézard)

The Acetylene Laboratory IR Spectrum: New Quantitative Studies

D. Jaccquemart, N. Lacome, and L. Gomez

Université Pierre-et-Marie-Curie-Paris6, Laboratoire de Dynamique, Interactions et Réactivité, CNRS, UMR 7075, Case courrier 49, 4, place Jussieu, 75252 Paris Cedex 05, France

J.-Y. MANDIN

Université Pierre-et-Marie-Curie-Paris6, Laboratoire de Physique Moléculaire pour l'Atmosphère et l'Astrophysique, CNRS, UMR 7092, case courrier 76, 75252 Paris Cedex 05, France

Abstract. The acetylene molecule $^{12}C_2H_2$ shows numerous vibration – rotation bands throughout the IR spectrum. Vibrational levels of C_2H_2 are grouped into clusters almost regularly spaced every 700 cm^{-1}, from the fundamental ν_5 band, at 13.6 μm, up to the visible. Several IR spectral regions where C_2H_2 bands occur have been extensively studied in the past years, mainly in order to obtain absolute individual line intensities and to improve spectroscopic databases as HITRAN or GEISA. This quantitative spectroscopy work is performed with the aid of Fourier transform interferometers to obtain absorption spectra, and using a multispectrum fitting procedure to retrieve line parameters from these spectra. For usual applications, a semi-empirical model based on the Herman-Wallis factor is used to generate line lists dedicated to spectroscopic databases. This poster gives a summary of all the spectral regions studied for acetylene $^{12}C_2H_2$, pointing out the current state of the spectroscopic databases HITRAN/GEISA. Works in progress (around 1300 cm^{-1}) and in project (0 - 500 cm^{-1}) will also be presented. Data available in the literature, or obtained in the recent works, have been compiled to set up line lists usable for applications and dedicated to databases. On the whole the number of transitions is twice compared to the actual HITRAN 2004 database plus the 2007 updates and is ranging from 700 to 9600 cm^{-1}.

The acetylene molecule C_2H_2 is of astrophysical interest. This molecule shows numerous vibration-rotation bands throughout the IR spectrum, its vibrational levels being grouped into clusters, or polyads, almost regularly spaced every 700 cm^{-1}, from the fundamental ν_5 band, at 13.6 μm, up to the visible. Several of the spectral regions where C_2H_2 bands occur, have been extensively studied in the past years, mainly in order to obtain absolute line positions or intensities, and to improve spectroscopic databases as HITRAN (Rothman et al. 2009) or GEISA (Jacquinet-Husson et al. 2008). Such a quantitative spectroscopy work is performed with the aid of Fourier transform interferometers to obtain absorption spectra, and using a multispectrum fitting procedure to deduce line positions and intensities. Then, for usual applications, a semi-empirical model based on the Herman-Wallis factor can be used in most cases to calculate synthetic spectra.

Data available in the literature, or obtained in recent works, have been compiled to generate line lists dedicated to the databases. Figure 1 shows the improvement brought by this update compared with the current HITRAN 2004 database plus its 2007 updates: 11055 transitions are now available between 600 and 9900 cm^{-1} for $^{12}C_2H_2$, 6232 coming from the present update (Jacquemart et al. 2009a,b). They concern 115 bands occuring in 14 spectral regions.

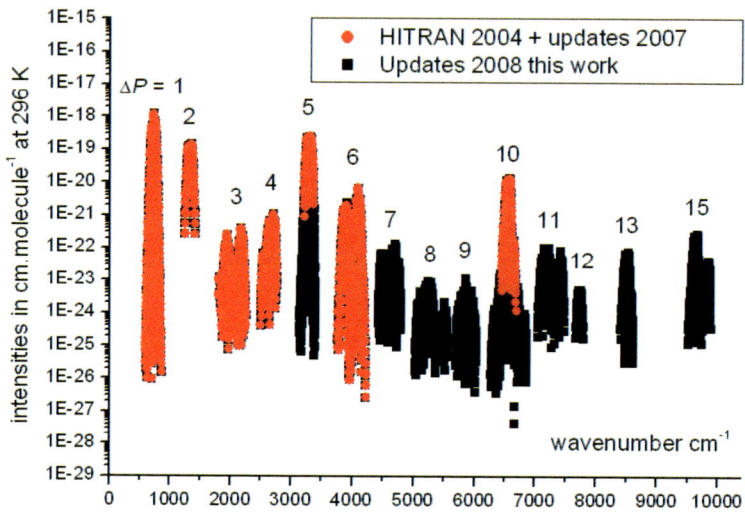

Figure 1. Intensities of transitions for $^{12}C_2H_2$ that will be available in databases. Each spectral region is named by its ΔP value, P being the pseudo-quantum number characterizing a polyad of interacting vibrational states.

Despite this improvement, databases are far from to be complete, as observable for example on Fig. 1 around 1300 cm^{-1}. In this spectral region, acetylene has been observed, e.g., in the circumstellar envelopes of carbon-rich stars. Using the Infrared Spectrograph (IRS) on board the Spitzer Space Telescope (SST), Matsuura et al. (2006) detected acetylene bands at 7 and 14 μm in carbon-rich asymptotic giant branch stars in the Large Magellanic Cloud. Around 7 μm, spectroscopic databases only contain line positions and intensities that Vander Auwera calculated from his absolute measurements in the $(\nu_4 + \nu_5)^0_+$ band (Vander-Auwera 2000), but intensities measured in Vander-Auwera (2000) for some lines of the $(\nu_4 + \nu_5)^2$ band are not reported in the databases. The temperature of interest for applications being around 500 K (Matsuura et al. 2006), the knowledge of the intensities in the numerous hot bands assigned by Kabbadj et al. (1991) is also important. In the quoted paper, Matsuura et al. (2006) could not reproduce the shapes that they observed in their IRS-SST spectra around 7 μm because of the lack of data in databases.

To complete the knowledge of C_2H_2 line intensities around 7 μm, a series of six Fourier transform spectra have been recorded with the rapid scan Bruker IFS 120 HR interferometer of the LADIR (Paris). The unapodized FHWM resolution is 2.8 x 10^{-3} cm^{-1} and the SNR about 350. The absorbing path length is 20 m and the temperature 296 K. Pressures from 0.2 to 7.6 hPa have been used. Absolute line intensity measurements have been performed with the aid of a multispectrum fitting procedure already detailled in previous works, see, e.g., Jacquemart et al. (2001). The accuracy of the measured intensities is 5% on the average.

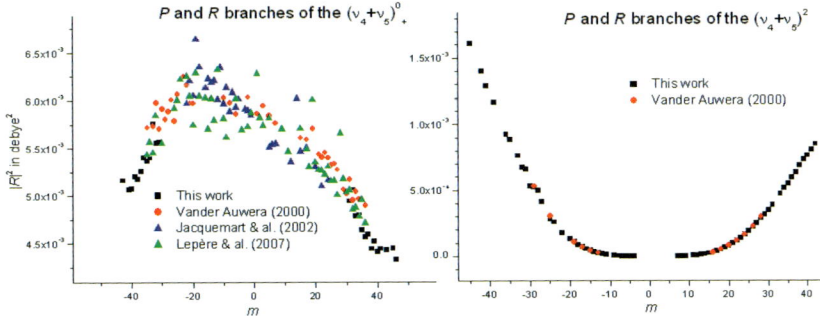

Figure 2. Variation of the transition dipole moment squared $|R|^2$, in D^2 (1 D = 3.33546 ×10^{-30} C.m) for the $(v_4 + v_5)^0_+$ and the $(v_4 + v_5)^2$ bands.

The spectra is very crowded (at least 1600 lines of 17 bands have been assigned by Kabbadj et al. (1991)), so that we present here the first results obtained in the two strongest bands: the $(v_4 + v_5)^0_+$ band and the "forbidden" $(v_4 + v_5)^2$ band, both studied by Vander-Auwera (2000). Figure 2 shows the rotational dependence of the transition dipole moments squared deduced from the measured line intensities (see Eq. 1 of Jacquemart et al. (2009a)), with $m = J$ in the P-branch and $J+1$ in the R-branch, J being the rotational quantum number. The first step of this quantitative spectroscopic work is to check carefully the consistency of existing measurements. For that, we calculated the average deviation between different set of measurements:

$$\left\langle \frac{\text{Lepere et al. (2007)} - \text{Vander Auwera (2000)}}{\text{Lepere et al. (2007)}} \right\rangle = 1.6(30)\% \text{ for 28 common lines,}$$

$$\left\langle \frac{\text{Jacquemart et al. (2002)} - \text{Vander Auwera (2000)}}{\text{Jacquemart et al. (2002)}} \right\rangle = 1.5(30)\% \text{ for 14 common lines,}$$

$$\left\langle \frac{\text{This work} - \text{Vander Auwera (2000)}}{\text{This work}} \right\rangle = 1.9(30)\%, \text{ for 22 common lines,}$$

with 1 standard deviation (SD) between parenthesis in unit of the last digit. These are in very acceptable agreement, with a small dispersion of the differences (1 SD 3%), if one takes into account the accuracies announced in the concerned works, i.e., 2% in Vander-Auwera (2000), and 5% on the average in Jacquemart et al. (2002), Lepère et al. (2007), and in the present work. Examination of Fig. 2 shows that the rotational dependence of the transition dipole moment squared of the $(v_4 + v_5)^0_+$ band will easily be modelled by the Herman-Wallis factor, despite the existence of resonances. On the contrary, first test performed on the $(v_4 + v_5)^2$ forbidden band have showed that the formula used for such bands failed to adjust the data. However, Fig. 2 shows that the precision of the measurements is enough to allow to smooth the obtained values, and to interpolate the missing lines that could not be measured. Thus, a line list in the HITRAN format, gathering data concerning the strongest bands around 7 μm, will be set up in a near future. A great amount of work has still to be made to measure the numerous remaining weak hot bands, in order to be able to perform a rigourous theoretical treatment as this based on the global model of Perevalov, Lobodenko, & Teffo (1997).

References

Rothman, L. S., Gordon, I. E., Barbe, A., Chris Benner, D., Bernath, P. F., Birk, M., Brown, L. R., Boudon, V., Champion, J. P., Chance, K., Coudert, L. H., Dana, V., Fally, S., Flaud, J. M., Gamache, R. R., Goldman, A., Jacquemart, D., Lacome, N., Mandin, J. Y., Massie, S. T., Mikhailenko, S., Orphal, J., Perevalov, V. I., Perrin, A., Rinsland, C. P., Simeckova, M., Smith, M. A. H., Tashkun, S., Tennyson, J., Toth, R. A., Vandaele, A. C., & Vander Auwera, J. 2009, J Quant Spectrosc Radiat Transfer, in press.

Jacquinet-Husson, N., Scott, N. A., Chédin, A., Crépeau, L., Armanta, R., Capelle, V., Orphal, J., Coustenis, A., Boone, C., Poulet-Crovisier, N., Barbe, A., Birk, M., Brown, L. R., Camy- Peyret, C., Claveau, C., Chance, K., Christidis, N., Clerbeaux, C., Coheur, P. F., Dana, V., Daumont, L., Debacker-Barilly, M. R., Di Lonardo, G., Flaud, J. M., Goldman, A. Hamdouni, A., Mikhailenko S., Hess, M., Hurley, M. D., Jacquemart, D., Kleiner, I., Kopke, P., Mandin, J. Y., Massie, S., Nemtchinov, V., Nikitin, A., Newnham, D., Perrin, A., Perevalov, V. I., Pinnock, S., Régalia-Jarlot, L., Rinsland, C. P., Rublev, A., Shreier, F., Schult, L., Smith, K. M., Tashkun, S. A., Teffo, J. L., Toth, R. A., Tyuterev, Vl. G., Vander Auwera, J., Varanasi, P., & Wagner, G. 2008, J Quant Spectrosc Radiat Transfer, 109, 1043

Jacquemart, D., Lacome, N., Mandin, J. Y., Dana, V., Tran, H., Gueye, F. K., Lyulin, O. M., Perevalov, V. I., & Régalia-Jarlot, L. 2009a, J Quant Spectrosc Radiat Transfer, in press.

Jacquemart, D., Lacome, N., & Mandin, J. Y. 2009b, J Quant Spectrosc Radiat Transfer, in press.

Matsuura, M., Wood, P. R., Sloan, G.C., & Zijlstra AA. 2006, Mon Not R Astron Soc, 371, 415

Vander-Auwera, J. 2000, J Mol Spectrosc, 201, 143

Kabbadj, Y., Herman, M., Di Lonardo, G., Fusina, L., & Johns, J. W. C. 1991, J Mol Spectrosc, 150, 535

Jacquemart, D., Mandin, J. Y., Dana, V., Picqué, N., & Guelachvili, G. 2001, Eur Phys J D, 14, 55

Jacquemart, D., Mandin, J. Y., Dana, V., Régalia-Jarlot, L., Thomas, X., & von der Heyden, P. 2002, J Quant Spectrosc Radiat Transfer, 75, 397

Lepère, M., Blanquet, G., Walrand, J., Bouanich, J. P., Herman, M., & Vander Auwera, J. 2007, J Mol Spectrosc, 242, 25

Perevalov, V. I., Lobodenko, E. I., & Teffo, J. L. 1997, SPIE, 3090, 143

Collisional Line Profiles of Na Perturbed by H$_2$

N. F. Allard

Observatoire de Paris, GEPI, UMR 8111, CNRS, 61 Avenue de l'Observatoire, F-75014 Paris, France

Abstract. The study of the shape of pressure broadened alkali resonance lines is of great practical and theoretical importance for brown dwarf atmospheres. Much effort has been expended in this direction, particularly in the last 5 years. These studies have emphasized the non-Lorentzian behavior of the line profiles. It is well known that the impact approximation, which assumes that collisions occur instantaneously, causes the Lorentz theory to fail not too far from the line center. Our aim here is to present unified calculations of sodium perturbed by molecular hydrogen emphasizing how they deviate from Lorentzian profiles. Calculations have been done for the $D1$ and $D2$ lines at 1000 K.

1. Introduction

A unified treatment of the shape of pressure-broadened alkali absorption line from near resonance to the far wing is obtained using autocorrelation formalism. This treatment includes finite duration of collision and requires the knowledge of molecular potentials, that is, the binary interaction between an alkali atom and a perturbing atom or molecule as a function of their separation. Additionally, the dependence of the radiative dipole moment on separation for each molecular state may significantly alter the far wing profiles from those computed with the usual assumption that the transition probability is invariant with separation. Complete details and the derivation of the theory are given by Allard et al. (1999).

2. Theory

The normalized profile $I(\omega)$ can be written as the Fourier transform of the dipole auto-correlation function $\Phi(s)$,

$$I(\Delta\omega) = \frac{1}{\pi} Re \int_0^{+\infty} \Phi(s) e^{-i\Delta\omega s} ds \qquad (1)$$

Here the correlation function Φ is not for a single isolated atom, but for an ensemble of sources each experiencing a different microscopic environment with a different temporal history during the radiative process.

$$\Phi(s) = e^{-n_p g(s)} \qquad (2)$$

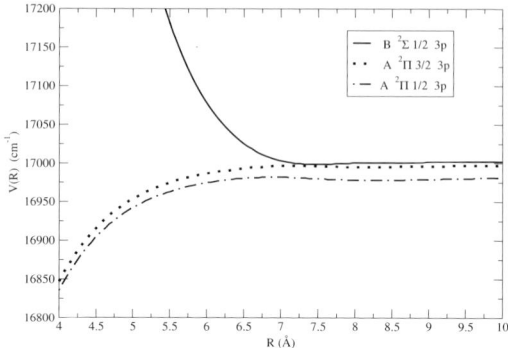

Figure 1. Na-H$_2$ potentials for the resonance lines for the C$_{\infty v}$ symmetry with spin-orbit coupling.

The well-developed theory of spectral line shapes allows us to compute the function Φ in which the density of perturbers is expressed explicitly, and the function $g(s)$ depends only on single collisions (Allard et al. 1999).

The decay of the autocorrelation function $\Phi(s)$ with time leads to atomic line broadening. It depends on the density of perturbing atoms n_p and on their interaction with the radiating atom. Fundamentally we are able to calculate a spectrum in this way for any neutral atom given the temperature, density, and composition of the gas in which it is found.

Moreover, the impact approximation determines the asymptotic behavior of the unified line shape correlation function. In this way the results described here are applicable to a more general line profile and opacity evaluation for the same perturbers at any given layer in the photosphere. We used the molecular-structure calculations performed by Rossi & Pascale (1985) for the molecular potentials of Na–H$_2$ system. For the specific study of the $D1$ ($P_{1/2}$) and $D2$ ($P_{3/2}$) components we need to take the spin-orbit coupling of the alkali into account. This is done using an atom-in-molecule intermediate spin-orbit coupling scheme, analogous to the one derived by Cohen & Schneider (1974). The degeneracy is partially split by the coupling and the distinction between $D1$ and $D2$ results.

Figure 1 presents the potential energy surfaces of Na-H$_2$ in the $C_{\infty v}$ symmetry. The shape of the line wing is sensitive to ΔV, the difference between the ground and excited state interaction potential. Satellites may appear on spectral lines when ΔV goes through extrema (Allard 1978).

3. Resulting Profiles

The line profiles have been calculated at $T = 1000$K for a fixed molecular hydrogen density ($n_{H_2} = 1 \times 10^{20}$ cm^{-3}). They are presented in Figs. 2-3 compared to the corresponding Lorentzian profile. The figures show a close-up of the 5800-6100 Å spectral region. The Lorentzian profiles are calculated using the line widths presented in Allard, Kielkopf, & Allard (2007). The $P_{1/2}$ line is due to a simple isolated $A \Pi_{1/2}$ state. The blue wings of the $D1$ line decrease very rapidly due to the shape of the potential of

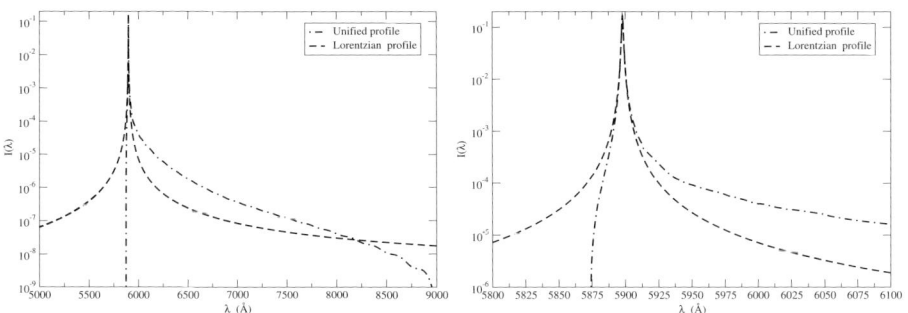

Figure 2. Na D1 line profile perturbed by H_2 for the $C_{\infty v}$ symmetry. The density of perturbers is $n_{H_2} = 10^{20}$ cm^{-3}. The temperature is 1000 K.

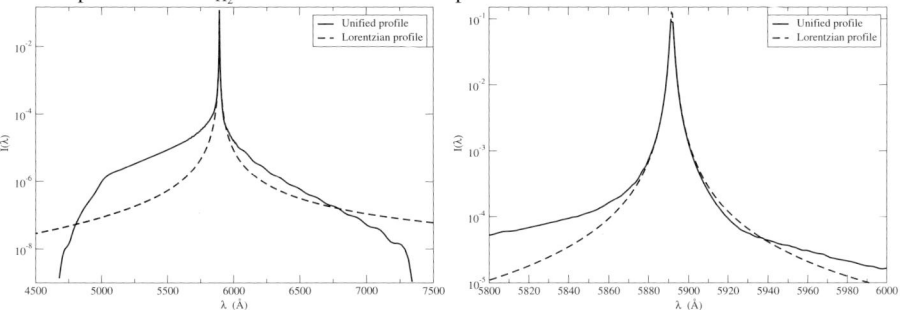

Figure 3. Na D2 line profile perturbed by H_2 for the $C_{\infty v}$ symmetry. The density of perturbers is $n_{H_2} = 10^{20}$ cm^{-3}. The temperature is 1000 K.

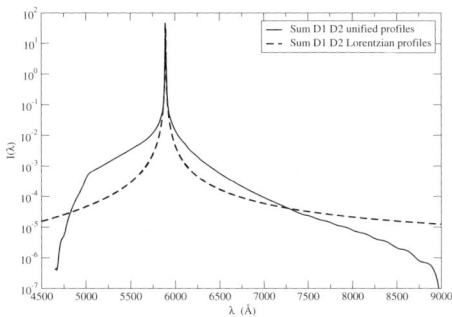

Figure 4. Na total unified line profile perturbed by H_2 (full line), compared to the corresponding Lorentzian profile (dashed lines). The density of perturbers is $n_{H_2} = 10^{20}$ cm^{-3}. The temperature is 1000 K.

the $A^2\Pi_{1/2}$ state, The line profile for the D1 line is then totally asymmetric (Fig. 2). Two excited molecular levels $B^2\Sigma_{1/2}$ and $A^2\Pi_{3/2}$ contribute to the D_2 line and produce opposite wings. While A states radiate on the red wing, B states radiate on the blue wing.

Blue satellite bands in alkali-He/H$_2$ profiles are correlated with maxima in the excited B state potentials and can be predicted from the maxima in the difference potentials ΔV for the B-X transition (Allard et al. 2003; Allard, Allard, & Kielkopf 2005; Allard & Spiegelman 2006; Allard, Spiegelman, & Kielkopf 2007), (Zhu, Babb, & Dalgarno 2005, 2006), (Alioua & Bouledroua 2006; Alioua et al. 2008). These line features arise from the changed structure of the radiating atom in the presence of another atom. As such, their presence is sensitive to the conditions of the radiating system and reveal the composition, temperature, and pressure of the source.

The total profile of the D$_1$ and D$_2$ lines of Na perturbed by collisions with H$_2$ is shown in Fig. 4.

Spectral line satellites have been seen in stars of extreme variety – on the Lyman series of hydrogen emitted by white dwarfs, and on alkali lines in brown dwarf stars (Allard, Kielkopf, & Loeillet 2004; Allard et al. 2007).

The line shape is now quantitatively understood, but the atomic interactions are difficult to calculate accurately. As a result, the science that can be achieved using spectral line shapes as a tool requires a continuing theoretical effort to establish the atomic interactions, and to implement that knowledge in predictive algorithms that can be used to extract new information from spectra.

References

Alioua, K. & Bouledroua, M. 2006, Phys. Rev. A, 74, 032711
Alioua, K., Bouledroua, M., Allouche, A. R. & Aubert-Frécon, M. 2008, J. Phys. B: At. Mol. Opt. Phys., 41, 175102
Allard, N. F., 1978, J. Phys. B: Atom. Molec. Phys., 11, 1383
Allard, N. F. , Royer, A., Kielkopf, J.F. & Feautrier, N. 1999, Phys. Rev. A, 60, 1021
Allard, N. F., Allard, F., Hauschildt, P. H., Kielkopf, J. F. & Machin L. 2003, A&A, 411, L473
Allard, N. F., Kielkopf, J. F. & Loeillet, B. 2004, A&A, 424, 347
Allard, N. F., Allard, F. & Kielkopf, J. F. 2005, A&A, 440, 1195
Allard, N. F. & Spiegelman, F. 2006, A&A, 452, 351
Allard, N. F., Kielkopf, J. F. & Allard, F. 2007 Eur. Phys. J. D, 44, 507
Allard, N. F. & Spiegelman, F. & Kielkopf, J. F. 2007, A&A, 465, 1085
Allard, F., Allard, N. F., Homeier, D., Kielkopf, J. F., McCaughrean, M. J. & Spiegelman, F. 2007, A&A, 474, L21
Cohen, J. S. & Schneider, B. 1974, J. Chem. Phys., 61, 3230
Rossi, F. & Pascale, J. 1985, Phys. Rev. A, 32, 2657
Zhu, C., Babb, J. F., & Dalgarno, A. 2005, Phys. Rev. A, 71, 052710
Zhu, C., Babb, J. F., & Dalgarno, A. 2006, Phys. Rev. A, 73, 012506

VUV Photophysics of Prebiotic Molecules in the Context of the Search for Life on exoplanets

S. Leach

Observatoire de Paris, LERMA, CNRS, 61 Avenue de l'Observatoire, F-75014 Paris, France

Abstract. The results of extensive studies of the VUV photophysics of a number of prebiotic molecules are discussed in the context of the survivability of these species under various conditions of irradiation and settings corresponding to the interstellar medium, planetary atmospheres and cometary and asteroidal environments. They are thus applicable to similar stellar conditions and planetary environments, and are also pertinent for identifying markers that can be associated with the presence or possible future existence of life on exoplanets.

1. Introduction

The search for exoplanets is intimately linked to the search for extraterrestrial life. Since the only life we know occurs in our solar system, it is logical to search for exoplanets in similar systems. No extrasolar systems with close similarities to our own have yet been discovered, due in part to observational bias, through lack of adequate technical means. Furthermore, present ideas as to the mechanisms of planet formation are in a state of flux. Thus the search for extrasolar systems similar to our own is an exciting and evolutionary pursuit. Our solar system contains not only planets but also comets, asteroids, meteorites, which could be sources for the molecular building blocks of life. In analogous circumstances they could penetrate whatever atmosphere exists on an exoplanet and thus deliver these key molecules to the atmosphere, the surface or any liquid haven. Material can also be transferred from one planet to another, as illustrated by the existence of Martian and Lunar meteorites on Earth. In addition, cosmochemistry in the interstellar medium (ISM) can be a source of prebiotic molecules that can eventually be deposited on planetary sites. The atmospheres of extrasolar planets will undoubtedly be studied by spectroscopic means. Direct, and other, more indirect, observations should enable the prevailing physical environment, in particular the local radiation field, to be determined. Thus it is imperative to understand the viability of prebiotic and biotic molecules in these sites. The results of extensive studies of the VUV photophysics of a number of prebiotic molecules are reported here. Our experimental results bear on the survivability of these species under various conditions of irradiation and settings corresponding to the interstellar medium, planetary atmospheres and cometary and asteroidal environments. They are thus applicable to similar stellar conditions and planetary environments, and are also pertinent for identifying markers that can be associated with the presence or possible future existence of life on exoplanets. The molecules discussed fall into two groups: 1) small molecules that are considered to be the reactants in a prebiotic chemistry which culminates in complex molecules such as amino acids and nucleobases, 2) the monomeric building blocks of biopolymers, e.g. aminoacids, the building blocks of proteins, and nucleobases which are part of the nucleotide building blocks of the informational molecules DNA and RNA. Our studies have involved both optical spectroscopy (absorption, fluorescence) and photophysical studies such as photoion mass spectrometry (PIMS), including the measurement of photoionization yields. Spectroscopic studies are essential for predicting observational possibilities in astronomy and for interpretation of laboratory photophysical results. The present report does not mention our absorption studies, which are published in detail elsewhere, and discusses only briefly the laboratory photophysics experiments whose details are also published elsewhere. Thus this report is limited to a presentation and discussion of some of the astrophysical implications of the photophysics results, more extensively for one of the small molecules, acetonitrile, and for the biologically important nitrogen heterocycles (pyrimidine, purine, imidazole and benzimidazole) as examples of what the photophysical results can offer in the context of the search for life in extrasolar systems.

2. Experimental

All the spectroscopic and photophysical experiments discussed here were done in collaboration with the group of Helmut Baumgartel, Free University, Berlin. They were carried out at synchrotron radiation sources BESSY I and BESSY II, Berlin, and LURE, Orsay. A typical example of the photophysics experiments is given here, that concerning acetonitrile. Monochromatised synchrotron radiation was obtained from the Berlin electron storage ring BESSY I (multi-bunch mode) in association with a 1.5 m normal incidence monochromator. For fluorescence measurements, the synchrotron light beam is focused into an open brass cell, differentially pumped, containing acetonitrile vapor at a pressure typically around 10^{-3} mbar. Fluorescence induced in irradiated target molecules passes through a quartz window and is dispersed using a 20 cm focal length secondary monochromator. The emitted fluorescence light, measured in the 250 - 700 nm wavelength range, is detected by a cooled photomultiplier. A dispersed fluorescence spectrum typically contains 2 points per nm. The resolution of these spectra is between 4 and 20 nm depending on the choice of the effective exit slit width. Dispersed fluorescence measurements were carried out at several excitation energies between 10 and 16 eV. The excitation bandwidth was ≈0.8 nm. In recording the fluorescence excitation (FEX) spectra, the secondary monochromator is fixed at a desired wavelength with a large exit slit and the primary monochromator is tuned in steps of typically 50-100 meV. The FEX spectra are corrected for the grating transmission function of the primary monochromator and for the VUV photon flux, respectively. The bandwidth was ≈ 0.3 nm, and the FEX spectral resolution was 40 meV at 13 eV photon excitation energy. Total ionization yields of acetonitrile were measured using a quadrupole mass spectrometer (Leybold Q200) at BESSY I. The ion yield curves were obtained through photon energy scans with measuring intervals of 10 meV. The measurement techniques used at BESSY I are essentially the same as those we used previously in a study of formic acid (fluorescence, Schwell, et al. 2002) and of PAHs (photoionization quantum yield, Jochims et al., 1996).

3. Prebiotic Molecules

3.1. Ammonia

Ammonia is one of the basic molecular species formed in the interstellar medium and in nebular condensations. Solar system comets and planetary atmospheres are also astrophysical sites of NH3. Spectral observation in extrasolar planets would be by infrared absorption. The VUV induced photodissociation of ammonia was studied by a dispersed fluorescence and fluorescence excitation techniques in the 6-16 eV excitation energy range (Leach et al. 2005). Ammonia itself fluoresces weakly, in the 220-250 nm region, but its photodissociation products NH2 and especially NH are strong emitters, respectively in the visible and in the near UV regions, and are good markers of ammonia, particularly in comets. Under certain physical and chemical conditions, ammonia, a nitrogen bearing species, can give rise to prebiotic molecules via reactions occurring in VUV processed ices containing NH3 (Cottin et al.1999).

3.2. Formic Acid

Formic acid has been observed in several sites in the ISM and appears to be a general component of ices occurring in the vicinity of embedded high-mass stellar objects. It is also observed in comets. The relative abundance of formic acid is at least 50 times greater in protostellar ices than in gas phase astronomical sources. This suggests that the gas phase photostability of formic acid with respect to UV and VUV radiation is much less for HCOOH embedded in a solid ice. Formic acid is also observed in high yield in Miller-Urey type experiments which simulate the evolution of Early Earth atmospheres (Miller 1998). Photofragment fluorescence spectroscopy was used to study the VUV photofragmentation of formic acid in the 6-23 eV region (Schwell et al. 2002), leading to competitive pathways forming excited OH and HCOO radicals, which should be taken into account in astrochemical schemes. These radicals can be considered as possible markers for formic acid. The quantum yield of ionization was also measured as a function of incident photon energy. Ionic dissociation processes become more important between 13.6 and 18 eV but neutral dissociation processes with large quantum yields still occur. The latter constitute the most efficient fragmentation pathways of formic acid below the HI astrophysical limit of 13.6 eV.

3.3. Acetic Acid

Acetic acid is observed in the ISM and is expected to occur in comets where, in both cases, VUV radiation is present. It is formed, as is formic acid, in Miller-Urey type experiments on simulated atmospheres of the Early Earth (Miller 1998). VUV photodissociation of gaseous acetic acid was studied in the 6-23 eV region, using synchrotron radiation excitation, photofragment fluorescence spectroscopy and photoion mass spectrometry (PIMS, Leach et al. 2006). The results show that in HI regions of the ISM parent ions will be formed above 10.6 eV, the yield being about 50% at 13.6 eV. Dissociative ionization occurs at photon energies above 11.6 eV. Protection from the destructive effects of VUV radiation is thus necessary to achieve detectable amounts of acetic acid in the ISM. This molecule has been observed in hot molecular cores, which are warm condensations inside molecular clouds associated with active star formation. The density of these condensations is such that they will offer some protection against radiation. A typical source is the molecular cloud complex Sgr B3, the acetic acid being found in a very small core, about 0.1 pc across. This polyatomic molecule is probably synthesized on dust grain surfaces and ejected into the ISM by evaporation or by shocks. It is known that ammonia and acetic acid can combine, in the laboratory, to produce aminoacids, in particular, glycine. This could possibly occur in the ISM but is more easily achieved in cometary and meteoritic material contexts, which can provide various catalytic media and radiation screening. Acetic acid is found in many meteorites. It is the most abundant of the monocarboxylic acids in the Murchison meteorite. These species may be the most unaltered interstellar molecules.

3.4. Acetonitrile

VUV induced dissociation of gaseous acetonitrile was studied in the 7-22 eV range using synchrotron radiation excitation, photofragment fluorescence spectroscopy and photoion mass spectrometry (Schwell et al. 2008). The dissociative photoionization onset, at 13.94 eV, is above the HI limit in the ISM, 13.6 eV. At this limit energy the photoionization quantum yield of acetonitrile is 0.53. Taking into account our measured VUV absorption cross sections, a value of $2.7 \times 10^{-9} s^{-1}$ is obtained for the integrated photodissociation rate of acetonitrile in the 115-180 nm region due to the interstellar field in the solar neighbourhood. Protection from the destructive effects of VUV radiation is necessary in the ISM. Neutral acetonitrile has been observed towards protostars in particular in hot molecular cores. Another source of CH_3CN is in protoplanetary discs around young stars, where most of the radiation flux is Lyman alpha. The stellar radiation field can differ considerably from that of the standard interstellar radiation field, especially for the cooler T Tauri stars. When the radiation flux is high, the photodissociation of the molecular ions, of which little is known experimentally, becomes important and should be included in model cosmochemical schemes. It should be remarked that, besides direct ionization events, a UV field can occur via the emission of X-rays by young stars, due to the formation of secondary electrons and their interaction with molecular hydrogen, giving rise to H2 VUV emission (van Dishoek et al. 2006). It is also important to note that the H Lyman-α flux at 121.5 nm (10.2 eV) can be extremely high in young stars. This is below the ionization energy, 12.2 eV, of acetonitrile. A study of the lifetime of acetonitrile during UV photolysis, related to its survival in space, has been made by Bernstein et al (2004). They irradiated a film of CH_3CN at 15 K, as well as in argon and water matrices at 15 K, with a hydrogen lamp producing photons essentially between 7.75 and 10.2 eV. From a measure of the rate of disappearance of the acetonitrile I.R. absorption under these conditions, they estimated the half-life of acetonitrile in the diffuse interstellar medium ($\approx$ 1000 years) and in solar system ices (weeks on upper surface of Europa). These values relate to integrated photolysis effects over only the 7.75-10.2 eV excitation range. They do not take into account the effects of VUV radiation below 10.2 eV which would include ionization at and above 12.201 eV. Furthermore, they are certainly dependent on solid state effects and on the chemical nature of the matrix. It is thus somewhat illusory to extrapolate the acetonitrile photodestruction rates of Bernstein et al. (2004) to the gas phase relevant in cometary, atmospheric and interstellar sites. Acetonitrile has also been observed in comets both by radiofrequency and mass spectroscopies. The fluorescence of its dissociation product, CN, is commonly observed early as a comet approaches the sun. The photodissociation rate of acetonitrile is required for calculations of acetonitrile formation and destruction in the stratosphere, in the atmosphere of Titan and in the ISM. Our measured value of 1390 Mb for the integrated absorption over the 106-180 nm region, has enabled us to determine the integrated photodissociation rate in these various astrophysical sites (Schwell et al. 2008a).

4. Biological Building Blocks

4.1. Aminoacids

A photoionization mass spectrometry study in the 6-22 eV photon energy region of five aminoacids, glycine, α-alanine, β-alanine, α-aminoisobutyric acid and α-valine, revealed a large number of VUV-induced degradation pathways of these biological molecules [10]. Ion pair formation appears to occur in some of the low-energy dissociation processes. Isomeric interconversion between α-alanine and β-alanine does not occur up to 20 eV excitation energy. Three of the aminoacids studied, glycine, α-alanine and α-valine, are proteinaceous, while all five have been found in meteoritic materials. The photoabsorption cross section of the amino acids is much higher in the VUV than in the UV. In particular, all of these molecules absorb strongly at 10.2 eV, where the Lyman-α stellar emission is intense. All five aminoacids would be easily photodissociated in HI regions of the ISM since their neutral and parent ion dissociation limits are well below 13.6 eV. Thus their survival in astrophysical sites under UV and VUV irradiation requires conditions of radiation shielding, such as occurs in dark interstellar clouds and in meteorites and micrometeorites. The nonchiral α-aminoisobutyric acid is related to chiral α-alanine via methyl group replacement of the hydrogen atom attached to the α-carbon of the latter. This non-chiral molecule is one of the most abundant aminoacids found in the Murchison meteorite. It has also been observed in micrometeorites recovered from the Antarctic (Maurette 1998).

4.2. Nucleobases: Adenine, Thymine and Uracil

Photoionization mass spectrometry in the 6-22 eV photon energy region was used to study the fragmentation patterns, ionization energies and fragment ion appearance energies of the nucleic acid bases adenine, thymine and uracil (Jochims et al. 2005). The fragmentation information is pertinent to understanding radiation damage in DNA and RNA. For the purine, adenine, the ion fragmentation is mainly governed by successive loss of HCN units. The two pyrimidines, thymine and uracil, have similar dissociation pathways, with main neutral loss pathways which involve HNCO and CO and, in uracil, HCN. The astrophysically important fragment ion, HCNH+ can be formed by several fragmentation pathways in all three nucleobases. These nucleobases would be easily photodissociated in HI regions of the ISM. Their observation in the ISM calls for study of regions of radiation protection (dark clouds) or regions where nucleobase production could successfully outweigh destruction. The existence of nucleobases in meteorites (Murchison, Murray and Orgueil) and micrometeorites also implies that their formation and survival occurs in conditions of efficient VUV radiation shielding. Laboratory work has suggested that purines and pyrimidines can be formed in the ISM by various chemical pathways (see below). Once formed in the ISM, these nucleobases, along with aminoacids, could be delivered to exoplanets in ways comparable to those invoked for solar system planets.

4.3. Pyrimidine, Purine, Imidazole and Benzimidazole

Purine is a molecular skeletal building block of the nucleic acid bases adenine and guanine, while cytosine, thymine and uracil are related to pyrimidine. Photoionization mass spectrometry in the 7-18 eV photon energy region was used to study the ionization processes and fragmentation pathways, ionization energies and fragment ion appearance energies of four nitrogen heterocyclic molecules, pyrimidine, purine, imidazole and benzimidazole, that are considered to be possible building blocks of life (Schwell et al. 2008b) Let us consider what could be the fate of the nitrogen heterocycles studied if they were formed in the interstellar medium (ISM). The ionization energies are respectively 9.21 eV (pyrimidine), 9.35 eV eV (purine), 8.66 eV (imidazole) and 8.20 eV (benzimidazole). A rule of thumb (Jochims et al. 1996) indicates that for a number of aromatic molecules of similar size to those studied here, the quantum yield of photoionization increases roughly linearly from the ionization onset and reached unity at about 9.2 eV above this onset energy. In HI regions in space, the upper energy limit of UV radiation is 13.6 eV. Thus, in these regions, the maximum ionization yields would be expected to be in order of magnitude: pyrimidine 48%, purine 46%, imidazole 54%, benzimidazole 59%. From our PIMS results, the lowest energy thresholds for dissociative ionization are: pyrimidine (ge11 eV), 12.27 eV; purine 12.6 eV, imidazole 11.34 eV, benzimidazole 13.2 eV. Thus, the four species studied could undergo dissociative ionization in HI regions, but with relatively small yields for purine and especially benzimidazole, whose dissociative ionization thresholds are 1 eV or less below 13.6 eV. We note that a study of the lifetime of pyrimidine under UV photolysis, related to its survival in space, has been made by Peeters et al (2005). They irradiated an argon matrix containing pyrimidine at 12 K with a hydrogen lamp emitting UV photons with a flux of 4.6×10^{14} photons $cm^{-2}s^{-1}$. From a measure of the rate of disappearance of the pyrimidine I.R. absorption under these conditions, they estimated the photodestruction cross section of pyrimidine (1.2×10^{-17} $cm^2 molecule^{-1}$), as well as the half-life, 8.1 yr, in the diffuse interstellar medium, and 0.81 Myr in dense interstellar clouds. These estimated values relate to integrated photolysis effects only up to Lyman-α 10.2 eV since there is relatively little emission above this energy by a microwave-excited hydrogen flow source, apart from the question of window transmission (not detailed). Thus in the work of Peeters et al no account is taken of the effects of VUV radiation above 10.2 eV. At this limiting energy, which is below the pyrimidine dissociative ionization energy, we estimate the ionization yield (in the gas phase) to be only 11%. We remark that the destruction rates and half-lives determined by Peeters et al are certainly dependent on solid state effects and on the chemical nature of the matrix (Cottin et al. 2003). It is thus somewhat illusory to extrapolate these rates to the gas phase relevant in interstellar sites. Laboratory work has suggested that purines and pyrimidines might be formed in the ISM starting from hydrogen cyanide or cyanoacetylene (Basile et al. 1984), or from even more simple gases like NH_3, N_2, CO_2 and H_2O (Lavrentiev et al. 1984). Speculation has been made on the possible formation of adenine in the ISM by successive condensation reactions of HCN (Chakrabati and Chakrabati, 2000) but this proposal has been contested by Smith et al (2001). A very recent computational study of mechanisms of prebiotic pyrimidine-ring formation of monocyclic HCN-pentamers, with application to adenine synthesis is space, has been carried out by Glaser et al. (2007). Observation of pyrimidine and imidazole in the interstellar medium (ISM) by radioastronomy has been attempted but has not been successful so far. Upper limits of abundance have been established for pyrimidine (Kuan et al. 2003) and imidazole

(Irvine et al. 1981). More recently, a particular interstellar spectral feature, previously observed in hot molecular cores towards W51 (Kuan et al. 2003), is considered as being possibly due to pyrimidine, and has been re-analyzed favorably (Charnley et al. 2005), but this is insufficient to confirm the pyrimidine assignment without justification via other lines, possibly due to pyrimidine, which require further observation and study. A search for pyrimidine in circumstellar envelopes of carbon-rich stars, thought to be a suitable site for pyrimidine observation, was unsuccessful (Kuan et al. 2003). Protection from the destructive effects of VUV radiation is necessary for detectable amounts of these nitrogen heterocyclic molecules to exist in the ISM. Observations should be fruitful towards protostars, in particular in hot molecular cores (Araya et al. 2005), which are warm condensations inside molecular clouds associated with active star formation regions, whose density is such that they will offer molecules some protection against radiation. Another interesting site to consider is the planet formation region of young circumstellar disks. It has been found that the abundances of simple organic molecules and water for AA Tauri, a typical young star with an accretion disk, are about one order of magnitude higher than for observed molecular cores (Carr and Najita 2008), indicating that molecular synthesis takes place within the disk. We suggest that in further attempts to observe these species, as well as nucleobases such as adenine, a suitable tracer could be the HCN molecule, the principal neutral fragment of dissociative ionization of these molecules, and a well observed species in hot cores in the ISM as well as being a species with a relatively high abundance in AA Tauri. The dissociative photoionization threshold energies for the ejection of HCN are: pyrimidine 12.27 eV, purine 12.6 eV, imidazole 11.41 eV, benzimidazole 13.2 eV (Schwell et al. 2008b), adenine 11.56 eV (Jochims et al. 2005). In the same spirit, we remark that HNCO would be a suitable tracer for thymine and for uracil (Miller 1998). Although pyrimidine is a precursor of thymine and uracil, the presence of O atoms in these nucleobases gives rise to important dissociative ionization channels involving loss of HNCO rather than HCN. HNCO has also been observed in hot cores in the ISM by radioastronomy. Turning to the solar system, we note that pyrimidines and purines have been reported in the data obtained with the PUMA dust impact mass spectrometer during the flyby of comet Halley by the Soviet spacecraft VEGA 1 (Kissel and Krueger, 1987). However, the mass resolution was insufficient to specify particular species in these molecular categories. The organic gas composition was measured aboard the GIOTTO spacecraft by the PICCA ion mass spectrometer which measured ions in the m/z = 10-210 range with an effective mass resolution, outside the ionopause, of about 3 daltons. Combining the results with other observations led to the conclusion that the molecular gas phase ions observed, which include relatively small ions containing nitrogen, are derived from larger polymers present in the cometary matter (Krueger et al. 1991). More recently, pyrolysis-GC-MS has been applied to analyse HCN-polymers which may be among the organic macromolecules most readily formed within the solar system (Minard et al. 1998). These polymers can form spontaneously from HCN in the presence of trace quantities of base catalysts. The resulting black solid contains the purine based nucleic acid bases adenine and xanthine (as methylated derivatives), and other nitrogen heterocyclic molecules. Prebiotic synthesis of adenine has been achieved under Europa-like conditions in a study of dilute solutions of NH_4CN frozen for 25 years at low temperatures (Levy et al., 2000). It has been shown that that the nucleobase uracil can be formed from a simulated primitive atmosphere of CO, N2 and H2O irradiated by protons (Kobayashi and Tsuji 1997). Purine based nucleic acid bases, as well as certain hydroxylpyrimidines, have been found in the formic acid extract of carbonaceous chondrites (Hua et al. 1986). Purines, and smaller amounts of pyrimidines, have been identified in meteorites (Schwartz 2002, Martins et al. 2008). Our results show that naked exposure to the VUV radiation emitted by the sun could lead to dissociative photoionization of purines and pyrimidines at energies greater than about 11 eV (wavelengths less than 112.7 nm). The flux of solar radiation in this spectral region was much higher in the early life of the Sun than it is today, with perhaps as much as 10^4 times more UV radiation below 200 nm (Zahnle and Walker, 1984). Thus protection provided by encapsulation of these prebiotic molecules in meteors, asteroids and comet nuclei is required for their survival in the solar system, whether they are formed early or late in the life of the Sun, before eventually reaching the Earth.

5. Conclusion

There has been a rapid development of observations of exoplanets. Recently the capacity to determine the nature of gases in exoplanetary atmospheres and measure other properties of these atmospheres has been demonstrated. This encourages the expectation that future observations will extend to terrestrial type exoplanets and other bodies with could harbour life or be suitable for its emergence. We can consider the solar system as a model for searching for such extraterrestrial sites since at present it is the only stellar system known to contain life. Life on Earth probably comes from one source since all its forms are based on DNA. The manner in which life evolved is reasonably well understood but the manner in which it emerged on Earth remains a matter of suppositions. One approach involves an influx of the building blocks of life delivered via impacts of extraterrestrial bodies that are as diverse as micrometeorites, meteorites, comets, asteroids and planets. The building blocks generally considered include prebiotic simple gases whose reactions in a terrestrial environment can lead to the formation of aminoacids and other biologically important species. They also include more complex molecules, already formed extraterrestrially, such as aminoacids, purines and pyrimidines, that have biological functions. We consider that the results of our studies of the VUV photophysics of these two classes of molecules are applicable to stellar conditions and planetary environments similar to those of the Solar system. These results can aid in the observation and identification of prebiotic species in exoplanet atmospheres as spectroscopic and other investigations develop. They also bear on the survivability of these species under various conditions of irradiation and settings corresponding to the interstellar medium, planetary atmospheres and cometary and asteroidal environments as observed within and from the Solar system. These conditions can be considered as models that require refinement through observations of exoplanet properties in order for their direct application to exoplanetary sites.

References

Araya, E., et al., 2005, ApJS, 157, 279
Basile B., Lazcano A.& Oro J., 1984, Adv. Space Res., 4, 125
Bernstein M.P., S.F.M.Ashbourn, S.A.Sandford & L.J.Allamandola, ApJ, 601, 365
Carr, J.S. & Najita, J.R., 2008, Science, 319, 1504
Chakrabati S & Chakrabati S.K., 2000, A&A, 369, L6
Charnley, S.B., et al. 2005, Adv.Space Res., 36, 137
Cottin H., Moore M.H. & BÃÏ'nilan Y., ApJ, 590, 874
van Dishoek E.F., Jonkheid B. & van Hemert M.C., 2006, Faraday Discuss., 133, 231

Glaser R., Hodgen B., Farrelly D., McKee E., 2007, Astrobiology 7, 455
Hua, L.L., et al. 1986, Origins of Life 16, 226
Irvine, W.M., Ellder, J., & Hjalmarson, A., et al., 1981, A&A, 97, 192
Jochims, H.W., Baumgartel, H. & Leach, S., 1996, A&A, 314, 1003
Jochims, H.W., et al., 2004, Chem. Phys., 298, 279
Jochims, H.W., Schwell, M., Baumgartel, H., Leach S., 2005, Chem. Phys., 314, 263
Kissel, J. & Krueger, F.R., 1987, Nature, 326, 755
Kobayashi, K., & Tsuji, T., 1997, Chem.Lett., 903
Krueger, F.R., Korth, A., & Kissel, J., 1991, Space Sci.Rev., 56, 167
Kuan, Y.J., Yan, C.H., Charnley, S.B., Kisiel, Z., Ehrenfreund, P. & Huang, H.C., 2003, MNRAS345, 650
Lavrentiev, G.A., Strigunkova, T.F. & Egorov, I.A., 1984, Origins of Life 14, 205
Leach, S., Jochims, H.W. & Baumgartel, H., 2005, Phys. Chem. Chem. Phys., 7, 900
Leach, S., Schwell, M., Jochims, H.W. & Baumgartel, H., 2006, Chem. Phys., 321, 171
Levy, M., Miller, S.L., Brinton, K., Bada, J.L., 2000, Icarus, 145, 609
Martins, Z., et al. 2008, Earth Planet.Sci.Lett., 270, 130
Maurette, M., 1998, in The Molecular Origins of Life, Ed. A.Brack, Cambridge Univ.Press, 147
Miller, S.L. 1998, in The Molecular Origins of Life, Ed. A.Brack, Cambridge Univ.Press, 59
Minard, R.D., Hatcher, P.G., Gourley, R.C., Matthews, C.N., 1998, Origins Life Evol. Biosphere, 28, 461
Peeters, Z., Botta, O., Charnley, S.B., Kisiel, Z., Kuan, Y.J. & Ehrenfreund, P., A&A, 433, 58
Schwartz, A.W., 2002, Origins Life Evol. Biosphere, 32, 395
Schwell, M., et al., 2002, J.Phys.Chem. A, 106 10908
Schwell, M., Jochims, H.W. & Leach, S. 2008a, Chem. Phys., 344, 164
Schwell, M., Jochims, H.W., Baumgartel, H., Leach S. 2008b, Chem.Phys., 353, 145
Smith, I.W.M., Talbi, D. & Herbst, E., 2001, A&A, 369, 611
Zahnle, K.J., Walker, J.C.G., 1982, Rev.Geophys.Space Phys. 20, 280

Signatures of Water Clouds on Exoplanets: Numerical Simulations.

Theodora Karalidi and Daphne M. Stam

SRON–Netherlands Institute for Space Research, Sorbonnelaan 2, 3584CA, Utrecht, the Netherlands

Christoph U. Keller

Astronomical Institute, Princetonplein 5, NL-3584CC, Utrecht, the Netherlands

Abstract. Clouds are of crucial importance for a planetary climate, because they store atmospheric volatiles, and because they scatter and absorb incident starlight and absorb, emit and scatter thermal radiation. Consequently, the detection and characterization of clouds on a planet can provide us with a wealth of information on the conditions on the surface. Here, we present numerically simulated flux and polarization spectra, from 0.3 μm to 1.0 μm, of starlight reflected by Earth–like exoplanets that are covered by horizontally homogeneous water clouds, for different cloud altitudes and particle sizes. Our results show that the degree of polarization P is sensitive to the particle size, in particular at phase angles between 30° and ~ 50° and around 90°, and to the cloud top altitude, in particular at wavelengths between 0.35 μm and 0.7 μm. The information in P should be easier to retrieve than that in F.

1. Introduction

With the discovery of exoplanets, the quest for signs of their habitability started. Clouds have two roles in this effort: they are of crucial importance for a planetary climate, but they can also obstruct signs of biological activity such as the red edge of vegetation (Montañés-Rodríguez et al. 2006). The detection and characterization (coverage, altitude, optical thickness and particle microphysical properties) of clouds on a planet can provide us with a wealth of information on the conditions in the atmosphere and on the surface.

Here, we present numerical simulations of flux and degree of polarization spectra, from 0.3 μm to 1.0 μm, of starlight reflected by Earth–like exoplanets that are covered by horizontally homogeneous water clouds, for different cloud altitudes and particle sizes. Special attention is paid to the degree of polarization, because polarimetry appears to be a powerful tool for the detection of exoplanets (Stam et al. (2004) and references therein), and also for their characterization, since the planet's degree of polarization as a function of the wavelength and/or the planetary phase angle, is sensitive to the structure and composition of the planetary surface and atmosphere (Stam 2008).

2. Description and Characterization of Polarized Light

Integrated over the stellar disk, direct starlight can be considered to be unpolarized. Starlight that is reflected by a planet is, however, usually polarized. Light can fully be described by a vector: $\mathbf{F} = [F, Q, U, V]$. Here F is the total flux, Q and U describe the linearly polarized flux and V the circularly polarized flux (see e.g. Hansen & Travis (1974)). The dimensions of F, Q, U and V are in W m^{-2}s^{-1}.

For a planet that is mirror–symmetric to our reference plane (the plane through the centers of the planet, host star and the observer (Stam 2008)), U and V are equal to zero, and we can define the degree of polarization as $P = -Q/F$. For $P > 0$ (i.e. $Q < 0$), the light is polarized perpendicular to the reference plane, while for $P < 0$ (i.e. $Q > 0$) the light is polarized parallel to the reference plane.

3. Our Radiative Transfer Simulations

The atmosphere of our model planet consists of five locally plane–parallel, horizontally homogeneous layers, that contain gas and, optionally, cloud particles. The surface is flat and black. The atmospheric temperature and pressure profile is based on that of the Earth's atmosphere (McClatchey et al. 1972).

Clouds are modeled as horizontally homogeneous layers consisting of liquid water particles with a standard size distribution (Hansen & Travis 1974). We use two particle sizes: model A, with an effective radius $r_{\text{eff}} = 2.0$ μm and an effective variance $v_{\text{eff}} = 0.1$ (Stam 2008) and model B, with $r_{\text{eff}} = 6.0$ μm and $v_{\text{eff}} = 0.4$ (based on van Diedenhoven et al. (2007)). The scattering properties of our model particles are calculated using a Mie code (de Rooij & van der Stap 1984). All clouds have an optical thickness of $b = 2$ at 0.55 μm. Their b at other wavelengths is calculated with the Mie code. We use three different cloud altitudes: case h_1, with the cloud ranging from 0 km to 2 km, case h_2 with the cloud ranging from 2 km to 4 km and case h_3 with the cloud ranging from 4 km to 6 km.

We use a doubling–adding code combined with a disk-integration method (see Stam (2008)), that fully takes into account multiple scattering and polarization. Spectrally, we concentrate on the continuum and ignore absorption by gases (e.g. O_2, H_2O, O_2). We normalize the calculated reflected F such that at a phase angle $\alpha = 0°$ it equals the planet's geometric albedo (Stam 2008). Degree of polarization P does not require normalization since it is a relative measure.

4. Simulation Results

In Fig. 1, F and P of the starlight reflected by the model planet are plotted as functions of the planetary phase angle α and the wavelength λ for clouds consisting of the model A or the model B particles. The clouds range from 2 km to 4 km.

Although the absolute value of F appears to be sensitive to the particle type, the change of F with α shows little sensitivity. Only for $30° < \alpha < 40°$, and $\lambda \leq 0.7$ μm, the model B particles leave a trace of the first rainbow in F. The latter is more pronounced in strength and in spectral range in P for both particle models, although its strength appears to be particle size dependent and largest for the model B particles. The secondary rainbow is less pronounced and blends with the peak in P that appears

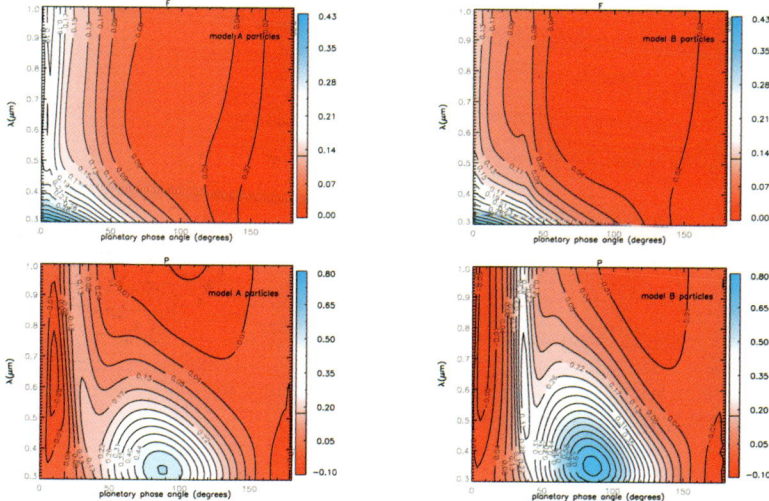

Figure 1. The reflected F (top) and P (bottom) as function of the phase angle and wavelength for the two different cloud particle types: model A (left) and model B (right). The clouds are have an optical thickness of $b = 2$ (at 0.55 μm) and their altitude ranges from 2 km to 4 km.

at short λ and around $\alpha = 90°$, and which is due to Rayleigh scattering from the gas molecules above the clouds.

Another factor that could influence the planetary signal is the cloud (top) altitude. In Fig. 2, F and P are plotted as functions of λ, for $\alpha = 90°$ and for a cloud consisting of model A particles placed at different altitudes. As can be seen in the figure, the maximum increase of F is only 3.4% (at $\lambda = 0.3$ μm), when the cloud top altitude decreases by 4 km. Such a small effect will be difficult to measure. On the other hand, P appears to be more sensitive to the cloud top altitude: it increases from 30% to ~ 40% (at $\lambda = 0.46$μm) when the cloud top altitude decreases from 6 km to 2 km. Whether or not this will be measurable will depend strongly on the design of the polarimetric system.

5. Summary

We have studied the effects of cloud particle size and cloud top altitude on F and P of starlight that is reflected by Earth–like exoplanets covered by horizontally homogeneous water clouds, as functions of the wavelength and planetary phase angle. Flux F appears to be relatively insensitive to both the particle size and the cloud top altitude: only with our largest particles ($r_{\text{eff}} = 6.0$ μm), the primary rainbow slightly increases F at phase angles around 40° and at wavelengths below 0.7 μm, and a cloud top altitude decrease of 4 km increases F at most 3.4% (at $\lambda = 0.3$ μm and $\alpha = 90°$), which will be very difficult to measure. The degree of polarization P appears to be more sensitive to the particle size and the cloud top altitude: the primary rainbow clearly shows, with a

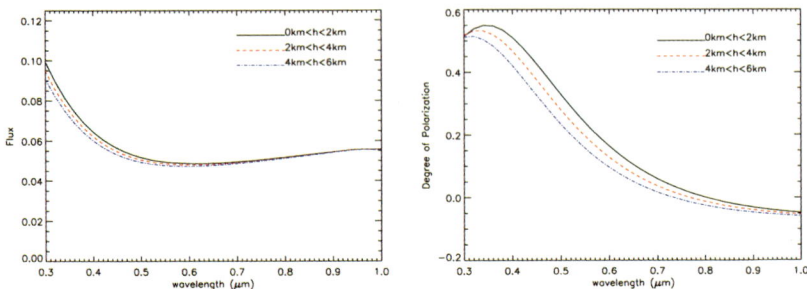

Figure 2. F and P as functions of the wavelength for clouds at different altitudes: between 0 and 2 km (black, solid line), between 2 and 4 km (red, dashed line) and between 4 and 6 km (blue, dot–dashed line). The cloud consists of model A particles and $b = 2$ at 0.55 μm. The phase angle is $\alpha = 90°$.

strength depending on the particle size, and P increases by 10% (at $\lambda = 0.46$ μm and $\alpha = 90°$) when the cloud top altitude decreases by 4 km.

Concluding, our simulations show that P is very sensitive to the sizes of the cloud particles and the altitude of the model clouds. This was also shown by e. g. Hansen & Hovenier (1974), who used ground–based polarization observations of Venus to derive the cloud properties. The dependence of P on the atmospheric composition and structure should be easier to measure than that of F.

Future work will involve a broader parameter study, testing the effect of more parameters (such as cloud optical thickness, surface albedo etc.) on the simulated signals, as well as expanding our research beyond the continuum to include also absorption by atmospheric gases. Simulations of more realistic planets, with mixture of surfaces (continents, sea etc.) and mixed atmospheres will follow.

References

van Diedenhoven, B., Hasekamp, O. P., & Landgraf, J. 2007, J. Geophys. Res. (A), 112, 15208
Hansen, J. E. & Hovenier, J. W. 1974, Journal of Atmospheric Sciences, 31, 1137
Hansen, J. E. & Travis, L. D. 1974, Space Sci. Rev., 16, 527
McClatchey, R. A., Fenn, R. W., Volz, F. E., & Garing, J. S. 1972, Optical Properties of the Atmosphere (AFCRL-72.0497))
Montañés-Rodríguez, P., Pallé, E., Goode, P. R., & Martín-Torres, F. J. 2006, ApJ, 651, 544
de Rooij, W. A. & van der Stap, C. C. A. H. 1984, A&A, 131, 237
Stam, D. M., Hovenier, J. W., & Waters, L. B. F. M. 2004, A&A, 428, 663
Stam, D. M. 2008, A&A, 482, 989

On the Protoplanetary-disk Origin of the Atmospheres of Hot Super-Earths

Masahiro Ikoma[1,2]

[1] *Department of Earth and Planetary Sciences, Tokyo Institute of Technology, Ookayama, Meguro-ku, Tokyo 152-8551, Japan.*

[2] *Observatoire de la Côte d'Azur, CNRS UMR 6202, BP 4229, 06304 Nice Cedex 4, France*

Abstract. While there is no such rocky planet in the solar system, super-Earths in extrasolar systems may have massive hydrogen atmospheres. In this paper I examine the possibility with results of my numerical simulations of the accumulation of the atmospheres of protoplanets embedded in protoplanetary disks. I also discuss how to distinguish between atmospheres of disk origin and chondritic origin.

1. Introduction

Recent development in detection techniques allows the discovery of low-mass exoplanets with mass of < 10 Earth masses. Especially observations of transiting planets will provide data of properties of their atmospheres. Thus it would be a good time to start exploring the origins and compositions of the atmospheres of Earth-like exoplanets.

A straightforward way would be to apply our understanding of the atmospheres of Earth and Venus directly to exoplanets' atmospheres. Indeed some recent papers address the issue in such an approach (e.g. Elkins-Tanton & Seager 2008). In the ancient solar system, possible volatile reservoirs were volatile-rich rocky planetesimals like carbonaceous chondrites, icy planetesimals like comets, and the solar nebula. The solar nebula is, however, unlikely to have made a major contribution to the Earth's atmosphere, because of the apparent difference in composition between the atmosphere and the Sun. A widespread idea is that volatile-rich planetesimals formed the atmosphere through degassing upon impact.

The idea is that volatile-rich planetesimals were delivered from outer parts of the solar system by the gravitational help of Jupiter. To form the atmospheres in that way, the gas giants must have formed from the solar nebula (i.e., the disk) first. Then, the disk gas disappeared before the terrestrial planets were fully formed; otherwise, nebular gas contributes greatly to their atmospheres. Also, if the disk gas had remained, it would have caused substantial migration of Jupiter.

However, analogs to our solar system are probably uncommon, as recently demonstrated by Thommes et al. (2008). Planets in general are formed in protoplanetary disks. Especially, transiting super-Earths are likely to have formed before the disks disappeared. This is because they should have undergone orbital migration that occurred via angular-momentum exchange between the planet and the protoplanetary disk. In this paper I examine the possibility of the protoplanetary-disk origin of the atmospheres of hot super-Earths and give constraints on their masses and compositions.

2. Accumulation of the Atmosphere

If a solid planet is embedded in a protoplanetary disk, it attracts the surrounding disk gas gravitationally to form an atmosphere. The atmosphere-disk boundary corresponds to the surface inside of which the planet's gravity dominates the gas' thermal motion. Since the boundary radius, namely, the total volume of the atmosphere is almost fixed, the mass of the atmosphere depends on its structure. The structure is determined by the thermal state of the atmosphere. For accreting atmospheres of this kind, the most important factor is the flux of heat supplied to the atmosphere; the atmospheric mass is known to increase almost linearly with decreasing heat flux (Ikoma & Genda 2006).

As far as hot super-Earths are concerned, they are likely to undergo giant collisions between planetary embryos. In that case a huge amount of energy is deposited in the interior of the planet, which delays the accumulation of the atmosphere. To evaluate the delay, I simulated the thermal evolution of a molten silicate sphere from extremely hot initial states. The numerical results show that the cooling occurs on the timescale of at most million years, which is short compared to the typical lifetime of a protoplanetary disk (Haisch et al. 2001).

Once the silicate part of the planet has cooled sufficiently, the atmosphere starts to cool down, resulting in accumulation of the atmosphere. Figure 1 shows the results of my numerical simulations of the evolution of the atmospheric mass for different four values of the planet's mass. Although a value of 10 Earth masses is often referred to as a critical value above which a planet have a massive hydrogen atmosphere, this figure shows that value is not necessarily critical. For example, even a 1-Earth-mass planet acquires an atmosphere of 1 % of the planet's mass, which corresponds to 10 times the total mass of oceans on the Earth. The 5-Earth-mass planet obtains more massive an atmosphere of several ten % of the planet's mass.

3. Implication for the Properties of the Atmosphere

The composition of a protoplanetary disk is similar to that of its parent star; it consists mainly of hydrogen and helium and contains small amounts of heavier elements. The atmosphere that came from the disk, however, has a different composition because of subsequent modification in composition.

A great part of the planet's mantle is molten because of the strong blanketing effect of the atmosphere. There would occur reactions between the atmosphere and molten mantle. One important reaction is addition of oxygen from the molten mantle to the atmosphere, because it produces water. As a matter of course, the opposite reaction could happen in the atmosphere, if the atmosphere is formed by the impact degassing of chondritic planetesimals (Elkins-Tanton & Seager 2008). Then, how can we distinguish between such two types of atmospheres?

One possible way is to measure the composition of the atmosphere. However, since water is rich in both atmospheres, water is not helpful to distinguish. The ratio of CO to H_2 or CH_4 to H_2, depending on temperature, would be helpful. According to Miller-Ricci et al. (2009), the thickness of atmospheres can be a good indicator to dis-

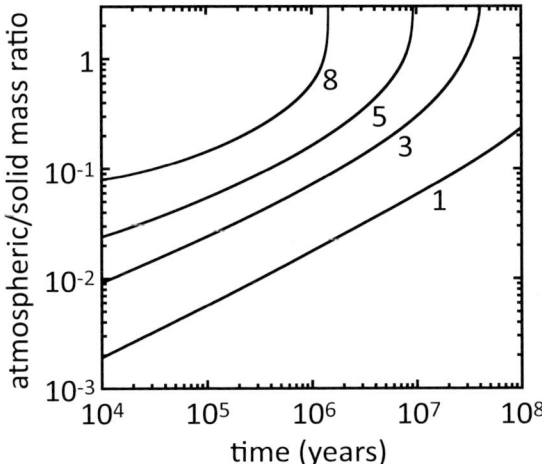

Figure 1. Increase in atmospheric mass due to accretion of gas from the protoplanetary disk. In those simulations a solid planet with constant density is assumed to be surrounded by an atmosphere with the solar composition. The vertical axis is the ratio of the atmospheric to solid mass. The number attached to each curve means the mass of the solid planet.

tinguish between hydrogen-rich and hydrogen-poor atmospheres, because the former is approximately 10 times as large as the latter. However, my equilibrium calculation concerning the atmosphere whose oxygen fugacity is buffered by the iron-wüstite buffer shows that the difference in mean molecular weight, namely scale-height, is at most a factor of 2 to 3, which might be too small to distinguish observationally.

When it comes to hot super-Earths, we have to keep in mind that all the hydrogen can be removed by intense stellar irradiation. In that case, there would be no good way to derive any information about the origin of the atmosphere from its current atmosphere. Even in that case, the relation between the planet's radius and mass (e.g. Valencia et al. 2007) could help us to distinguish between those two types of atmospheres. The atmosphere from a protoplanetary disk reduces the molten mantle and then produce metallic iron, which ends up growing the metallic core. Hot super-Earths that have large amounts of water and contain huge metallic cores will be possibly discovered.

Acknowledgments. This research was supported by the OPERA PPF and OCA BQR.

References

Miller-Ricci, E., Seager, S., & Sasselov, D. 2009, ApJ, 690, 1056
Elkins-Tanton, L. T., & Seager, S. 2008, ApJ, 685, 1237
Haisch, K. E., Jr., Lada, E. A., & Lada, C. J. 2001, ApJ, 553, L153
Ikoma, M., & Genda, H. 2006, ApJ, 648, 696
Thommes, E. W., Matsumura, S., & Rasio, F. A. 2008, Science, 321, 814
Valencia, D., Sasselov, D. D., & O'Connell, R. J. 2007, ApJ, 665, 1413

Cassini room was full for a special session where the first direct images of exoplanets were presented.

Part V

Brown Dwarfs
(Chair: Jean-Pierre Maillard)

The Brown Dwarf-Exoplanet Connection

Adam J. Burgasser

Massachusetts Institute of Technology, Dept. Physics Building 37-664B, 77 Massachusets Ave., Cambridge, MA 02139,USA

Abstract. Brown dwarfs are commonly regarded as easily-observed templates for exoplanet studies, with comparable masses, physical sizes and atmospheric properties. There is indeed considerable overlap in the photospheric temperatures of the coldest brown dwarfs (spectral classes L and T) and the hottest exoplanets. However, the properties and processes associated with brown dwarf and exoplanet atmospheres can differ significantly in detail; photospheric gas pressures, elemental abundance variations, processes associated with external driving sources, and evolutionary effects are all pertinent examples. In this contribution, I review some of the basic theoretical and empirical properties of the currently known population of brown dwarfs, and detail the similarities and differences between their visible atmospheres and those of extrasolar planets. I conclude with some specific results from brown dwarf studies that may prove relevant in future exoplanet observations.

1. A Brown Dwarf Primer

Brown dwarfs are stellar objects with insufficient mass to sustain core hydrogen fusion reactions, resulting in a steady decline in both luminosity and effective temperature (T_{eff}) with time. The mass limit for sustained hydrogen fusion is roughly 0.072 $M_\odot$ (75 Jupiter masses) for a Solar metallicity gas mixture, increasing to 0.090 $M_\odot$ for a pure hydrogen gas (e.g., Chabrier & Baraffe 2000). This mass limit establishes a formal division between "stars" and "brown dwarfs", although such a division is not necessarily relevant to how these objects form. While there is ongoing debate over the details of brown dwarf formation (the roles of gas turbulence, fragmentation and dynamical interactions; see recent reviews by Luhman et al. 2007 and Whitworth et al. 2007), observational evidence indicates brown dwarfs are created in a manner similar to, or at least coincident with, stars, via gravitational collapse of dense cores within giant molecular clouds. As a brown dwarf's energy reservoir arises primarily from the gravitational potential energy released in their initial contraction,[1] the luminosity, T_{eff} and emergent spectral energy distribution of a brown dwarf depend primarily on mass and age, and secondarily on elemental abundances, bulk properties (e.g., rotation) and external drivers (e.g., the presence of close companion). The interdependence of these factors on brown dwarf observables challenges the characterization of individual

---

[1] Small contributions also arise from brief periods of lithium- and deuterium fusion for objects more massive than ~0.065 $M_\odot$ and ~0.012 $M_\odot$, respectively. The latter limit is considered a possible dividing line between "brown dwarfs" and "planets" (see Basri & Brown 2006), an issue that will not be touched upon here.

sources in the well-mixed Galactic population; however, it also provides an opportunity to study a broad range of low-temperature atmospheric properties and processes.

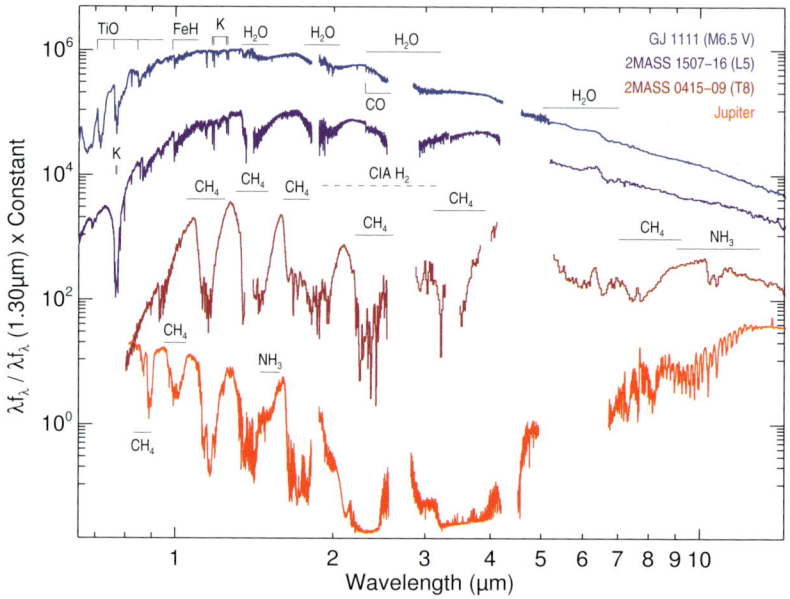

Figure 1. Observed optical to mid-infrared (0.65–14.5 μm) spectra of representative M-type, L-type, and T-type dwarfs, compared to data for Jupiter (top to bottom). Dwarf spectra are from Cushing et al. (2006) and references therein; Jupiter data are from Rayner et al. (2009) and Kunde et al. (2004). Spectra are arbitrarily normalized. Major molecular absorption bands characterizing these spectra are labeled, including TiO, FeH, H_2, H_2O, CO, CH_4 and NH_3. Atomic K I absorption is also labeled, which produces a substantial pressure-broadened line feature spanning 0.7–0.85 μm in L and T dwarf spectra. Note that Jupiter's emission shortward of $\sim$4 μm is dominated by scattered solar light modulated by CH_4 and NH_3 absorption features, while the dwarf spectra are entirely emergent flux (from Marley & Leggett 2008).

Brown dwarfs have been directly observed since the mid-1990s,[2] and there are now hundreds known to exist in young clusters, as companions to nearby stars, and, most commonly, as faint isolated systems within a few hundred parsecs of the Sun. The currently known population is segregated into three spectral classes based on the morphology of their optical or near-infrared spectra: M dwarfs, L dwarfs and T dwarfs (Figure 1). M dwarfs encompass the warmest, youngest, and most massive brown dwarfs which have had little time to cool. They exhibit spectral traits similar to older, low-mass dwarf stars, with strong metal-oxide molecular bands (including TiO, VO, CO

---

[2]On a historical note, both the discovery of the first widely-accepted brown dwarf, Gliese 229B (Nakajima et al. 1995), and the discovery of the first extrasolar gas giant planet, 51 Peg b (Mayor & Queloz 1995), were announced to the community in the same conference, Cool Stars 9, in October 1995; see Oppenheimer et al. (2000) for a historical review.

and H_2O) and neutral atomic line absorption blanketing their emergent spectral energy distributions. L dwarf spectra are characterized by strong metal-hydride (FeH, CrH), H_2O and CO molecular absorption; and alkali lines, including the heavily pressure-broadened Na I and K I doublets that largely sculpt the optical spectra of these sources (e.g., Allard et al. 2003; Burrows & Volobuyev 2003). L dwarfs also show evidence of condensate clouds in their photospheres, which give rise to highly reddened spectral energy distributions and absorption features from silicate grains (Cushing et al. 2006; see §3.1). T dwarfs are the coldest class of brown dwarfs currently known, characterized by H_2O, CH_4, NH_3 and strong collision-induced H_2 absorption. T dwarfs do not appear to have abundant condensate material in their photospheres. A fourth spectral class, the Y dwarfs, has been proposed for brown dwarfs even cooler than class T, although there is as yet no consensus on the general properties of this class nor a widely-accepted prototype (see Delorme et al. 2008; Burningham et al. 2008). The M, L and T spectral classes coincide roughly with T_{eff} ranges of $\gtrsim$ 2400 K, 2400 $\lesssim T_{eff} \lesssim$ 1400, and 1400 $\lesssim T_{eff} \lesssim$ 600, respectively (Golimowski et al. 2004; Vrba et al. 2004), although the end-point of the T spectral class remains uncertain. Variations in secondary parameters, such as metallicity, age and cloud properties modulate this temperature scale (e.g., Burgasser et al. 2006; Metchev & Hillenbrand 2006; Burgasser et al. 2008). For more information on the L and T spectral classes, see the recent review of Kirkpatrick (2005).

Molecules are a prominent feature of brown dwarf atmospheres and are fundamental in our ability to ascertain the physical properties of individual sources. Beyond spectral classification, the presence, relative strengths and detailed shapes of molecular features observed in brown dwarf spectra enable measures of T_{eff}, surface gravity, metallicity, cloud composition, atmospheric dynamics, rotation, and even the presence of unseen companions (e.g., Luhman 1999; Burgasser et al. 2006; Saumon et al. 2006; Burgasser et al. 2008; Cushing et al. 2008; Reiners & Basri 2008). Extracting these details for individual brown dwarfs is a current topic of interest in the field, and a challenge due to persistent inadequacies in theoretical spectral models and opacity line lists. The complex opacities of warm molecular gases and strongly pressure-broadened atomic features (e.g., Freedman et al. 2008; also see contribution by Tennyson), dynamical effects on gas chemistry (e.g., Griffith & Yelle 1999), and the complex processes associated with condensate grain formation (e.g., Ackerman & Marley 2001; Helling & Woitke 2006) are major hurdles in bringing atmospheric models into detailed agreement with observational data. Progress is being made on the theoretical front through new work on grain formation (e.g. Helling et al. 2008; see contributions by Allard and Freytag), quantum opacity calculations for key molecules (e.g., Barber et al. 2006), and incorporation of nonequilibrium chemistry (e.g., Saumon et al. 2006; see contribution by Homeier). On the observational side, the identification of benchmark sources—companions to age-dated stars, coeval cluster members, and resolved astrometric and eclipsing binaries—are a priority as critical tests of advanced models (e.g., Mohanty et al. 2004b; Zapatero Osorio et al. 2004; Leggett et al. 2008; Dupuy et al. 2008).

2. Comparing Exoplanets to Brown Dwarfs

The benefit of brown dwarfs to exoplanet studies lies in our current ability to study their atmospheres in considerable detail, over a broad range of wavelengths and spectral resolutions, and over time. Yet for brown dwarfs to be used as reliable templates for exoplanetary studies, it is essential to first assess whether their emergent spectra faithfully guide our interpretations of emergent/reflectance planetary spectra. To this end, I examine some of key similarities and differences in the physical properties and processes of brown dwarf and exoplanet atmospheres.

2.1. Temperatures

A gross assessment of the photospheric temperatures of brown dwarfs can be inferred from their T_{eff}s. These are typically determined from bolometric luminosity measurements and an assumed (theoretical) radius estimate (e.g., Golimowski et al. 2004; Vrba et al. 2004); alternately, fits of spectral data to theoretical models are used (e.g., Mohanty et al. 2004a; Burgasser et al. 2006; Cushing et al. 2008). These measures do not always agree (Smith et al. 2003). For planets, a comparable statistic is the thermal equilibrium temperature, $T_{eq} = T_*(R_*/2a)^{1/2}$, where T_* and R_* are the effective temperature and radius of the host star, respectively, and a the semi-major axis (ignoring albedo and orbital eccentricity). As it turns out, the T_{eff}s of L- and T-type brown dwarfs overlap considerably with the T_{eq}s of transiting extrasolar planets (Figure 2). Similarly, the directly-imaged planets HR 8799bcd (Marois et al. 2008), Fomalhaut b (Kalas et al. 2008) and β Pictoris b (Lagrange et al. 2008) have estimated T_{eff}s (not T_{eq}s; see below) comparable to T dwarfs. Fomalhaut b may in fact be cooler than the T_{eff} = 575±25 K ULAS 1335, the coldest brown dwarf currently known (Burningham et al. 2008).

Transiting planets are warm due to the radiative forcing by their host stars. A planet with T_{eq} = 500 K lies only 0.3 AU (0.07 AU) from a solar-type (M0 dwarf) primary. As $T_{eq} \propto T_* a^{-1/2}$, more widely-orbiting planets and planets orbiting less luminous host stars have lower T_{eq}s, below the range currently sampled by brown dwarfs. For closely-orbiting, tidally-locked hot Jupiter planets, care must be taken when using T_{eq} as a proxy for photospheric temperature, as these planets can have substantial day/night asymmetries (see contribution by Knutson). Eccentricity effects can also give rise to large temporal modulations in T_{eq} (see contributions by Iro and Lewis). In contrast, HR 8799bcd, Fomalhaut b, and β Pictoris b have (to first order) uniformly warm photospheres dominated by internal heat rather than reprocessed host star light. These planets are still young (<300 Myr); like brown dwarfs, their atmospheres will eventually cool to low temperatures.

2.2. Photospheric Pressures

While the mean photospheric gas temperatures of brown dwarfs and planets are comparable, gas pressures are generally quite different. At the photosphere, gas pressure is proportional to the surface gravity, g, as $P_{ph} \propto g/\kappa$, where κ is the Rosseland mean opacity. Evolutionary models dictate that the surface gravities of brown dwarfs depend strongly on mass (due to their nearly constant radii) and weakly on age (significant variations only for ages ≲ 100 Myr; see Figure 2). Surface gravities for evolved M, L and T dwarfs (ages ~0.5–10 Gyr) span g~300–3000 m s^{-2}. In contrast, the vast majority of transiting exoplanets have g~10–30 m s^{-2}, as directly inferred from radial velocity and transit light curves (e.g., Sozzetti et al. 2007). Ignoring opacity effects, the photospheric gas pressures of transiting exoplanets are 1–2 orders of magnitude less than those of brown dwarfs. Cooler, widely-orbiting Jupiter-mass gas giants also have photospheric pressures about 10 times less than their (typically more massive) brown dwarf counterparts.

Differences in photospheric gas pressure can have a measurable influence on some chemical pathways. One example is the reduction reaction $CO + 3H_2 \rightarrow CH_4 + H_2O$, which favors CH_4 production in high-pressure gas environments. Chemical equilibrium models indicate that CH_4 becomes abundant in brown dwarf photospheres (1-10 bar) below ~1400 K, but in planetary photospheres (0.01–1 bar) below ~900 K (e.g., Lodders & Fegley 2006). Pressure also modulates some gas opacities, notably the pressure-broadened alkali lines that dominate the optical spectra of L and T dwarfs, and collision-induced H_2 absorption that suppresses broad swaths of infrared light in the coldest brown dwarfs (e.g., Linsky 1969; Saumon et al. 1994). Both features are used to constrain surface gravities for individual brown dwarfs near the Sun (e.g., Martín et al. 1999; Burgasser et al. 2006; Kirkpatrick et al. 2006) and verify the membership of brown dwarfs in young clusters (e.g., Luhman 1999; Allers et al. 2007).

Fortuitously, there is overlap in photospheric gas pressure/temperature space between the youngest and lowest-mass brown dwarfs—those found in young star-forming regions and associations—and dense gas giant planets that are either very massive or have a substantial core. Transiting planets such as HD 147506b (aka Hat-P-2b; $\rho \approx 13$ g cm^{-3}; Bakos et al. 2007) and CoRoT-Exo-3b ($\rho \approx 26$ g cm^{-3}; Deleuil et al. 2008) have surface gravities similar to 30–100 Myr, ~5–20 Jupiter mass brown dwarfs like 2MASS 1207-39B and AB Pic B (Chauvin et al. 2004, 2005; Mohanty et al. 2007; see Figure 2). With a mass of 22 Jupiter masses, CoRoT-Exo-3b could be properly classified as a highly irradiated brown dwarf companion.

In addition to mean values of photospheric temperature and pressure, differences in the pressure-temperature profiles of brown dwarfs and exoplanets must be considered. For planets, external heating from the host star flattens out the pressure-temperature profile and can give rise to inversion layers. This translates into variations in the local gas chemistry and changes in the atmospheric column abundances of atomic and molecular absorbers. In addition, strongly irradiated planets develop deep radiative envelopes that extend well below the visible photosphere, whereas brown dwarf atmospheres are fully convective through to their photospheres (e.g., Burrows et al. 1997). Differences in the gas mixing rates and vertical temperature profiles between externally heated planetary atmospheres and brown dwarf atmospheres can produce profound differences in emergent spectral energy distributions, even for sources with comparable photospheric gas temperatures and pressures (e.g., Fortney et al. 2008a).

2.3. Compositions

The elemental composition, or metallicity, of a cool atmosphere also modulates chemistry and spectral appearance. Gas-giant planets tend to have metal-rich atmospheres, having condensed out of the gas-depleted debris disks around preferentially metal-enriched host stars (e.g., Gonzalez 1997). Ice-giant (i.e., Neptune) and terrestrial planet atmospheres exhibit even greater metallicity enhancements. These trends are present in the solar system: the atmospheres of Jupiter, Saturn, Uranus and Neptune have effective metal abundances ranging from ~3 to ~40 times that of the Sun (Fortney 2007). More importantly, there is a large range in individual elemental abundances, driven by the segregation of volatiles in the Sun's early protoplanetary disk and chemical separation in planetary atmospheres (e.g., He settling in Saturn). Significant variations in elemental abundances can have as great or greater impact on the chemistry and molecular composition of cool atmospheres as pressure or temperature alone (e.g., Tinetti et al. 2007; Fortney et al. 2008b).

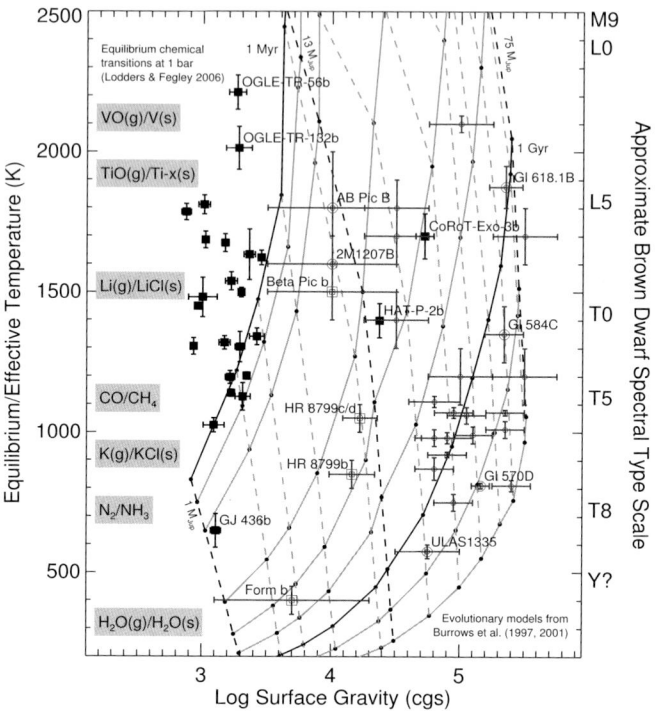

Figure 2. Atmospheric gas properties of extrasolar planets and brown dwarfs, as traced by T_{eff}/T_{eq} and surface gravity. T_{eq} and g values for transiting planets are from Torres et al. (2008) and indicated by solid black squares (outliers are specifically labeled). Inferred T_{eff} and g parameters for the directly detected planets, HR 8799bcd (Marois et al. 2008), Fomalhaut b (Kalas et al. 2008), and β Pictoris b (Lagrange et al. 2008) are indicated by open squares and labeled. T_{eff} and g parameters for several field brown dwarfs (Burgasser et al. 2006; Cushing et al. 2008) and a sample of benchmark sources (labelled; Mohanty et al. 2007; Kirkpatrick et al. 2001; Wilson et al. 2001; Saumon et al. 2006; Chauvin et al. 2005; Burningham et al. 2008) are indicated by open circles. These measurements are compared to the brown dwarf ("hot start") evolutionary models of Burrows et al. (1997, 2001). Solid lines delineate ages of 1, 5, 10, 30, 100, 300 Myr and 1, 3, and 10 Gyr, with 1 Myr and 1 Gyr isochrones highlighted. Dashed lines delineate masses of 0.001, 0.002, 0.003, 0.005, 0.01, 0.012, 0.02, 0.03, 0.04, 0.05, 0.06, 0.07 and 0.072 $M_\odot$, with the 0.001 $M_\odot$ = 1 Jupiter mass, 0.012 $M_\odot$ = 13 Jupiter masses (deuterium burning limit) and 0.072 $M_\odot$ = 75 Jupiter masses (hydrogen burning limit) lines highlighted. An approximate spectral classification scale based on T_{eff} determinations by Vrba et al. (2004) is indicated on the right axis. Equilibrium chemical transitions for several key species are indicated along the left side of the plot (Lodders & Fegley 2006).

In contrast, brown dwarf metallicites are expected to span the same range as stars, topping out at perhaps 3–5 times solar abundances but extending down to significantly subsolar abundances in the metal-poor thick disk and halo populations. For example, members of the recently-identified L subdwarf class have metallicities ~0.01-0.1 times solar (Burgasser et al. 2007; Schilbach et al. 2008). Even brown dwarfs with bulk solar metallicities will have slightly metal-poor photospheres due to condensation effects. Like stars, relative elemental abundances of brown dwarfs are likely to exhibit only small variations, although condensation effects may modify abundance patterns. In any case, the broad range of elemental abundances observed in the solar planets and expected to a greater degree among the wider exoplanet population will probably not be realized among Galactic brown dwarfs.

2.4. Stellar Hosts and Driving Forces

Unlike the majority of brown dwarfs, exoplanets are generally accompanied by a luminous host star, which ultimately maintains its atmosphere in a warm state, modulo variations arising from orbital eccentricity, circulation, or magnetic interaction effects. Radiation and stellar winds drive non-equilibrium dynamics in exoplanet atmospheres, including internal winds/jets (see contribution by Showman) and atmospheric stripping (see contribution by Alyward). UV and X-ray radiation drive photochemical production of hazes in exoplanet atmospheres (see contribution by Yung), and the formation of upper inversion layers. Tidal locking from a close stellar companion slows an exoplanet's rotation and can provide a source of (temporary) internal heating.

These processes will not generally occur in brown dwarf atmospheres. The hottest and youngest brown dwarfs do exhibit high-energy nonthermal emission (X-ray and UV) arising from magnetic activity or accretion. However, with the exception of rare massive flares in which the total magnetic energy output can briefly exceed thermal emission (e.g., Liebert et al. 1999), high-energy magnetic emission is typically a small fraction ($<10^{-3}$) of the total energy budget and does not significantly alter pressure-temperature profiles. Furthermore, for cooler L and T-type brown dwarfs, magnetic emission is conspicuously absent due to the loss of field/gas coupling in the highly neutral photospheres of these objects (Mohanty et al. 2002; Gelino et al. 2002). Only the very youngest (<1 Myr), actively accreting brown dwarfs and sources in close orbits around luminous companions are likely to show significant modification of their pressure-temperature profiles as a result of external driving forces.

The absence of both radiative and mechanical forcing on brown dwarf atmospheres is particularly relevant to atmospheric circulation and dynamics (see contributions by Showman and Cho). Rotational modulation of weather phenomena has been invoked to explain low-level spectral and photometric variability detected in some brown dwarfs (see review by Goldman 2005). The timescales for these variations are generally consistent with rotational line broadening measurements (e.g., Reiners & Basri 2008) and variations in nonthermal magnetic emission (e.g., Berger 2006). The general absence of magnetic field coupling (i.e., spots) and the clear presence of condensate clouds makes weather an appealing explanation for this variability, particularly given observed long-term period variations and changes in variability amplitudes. However, the winds and jets that drive weather in planetary atmospheres arise from asymmetric radiative forcing by the host star; such forces are absent for most brown dwarfs. It is possible that winds could be driven by the extremely rapid rotations of brown dwarfs. Periods of 1–10 hours are typical (cf. Jupiter's 11-hour period) and surface rotational velocities of up to 80 km s^{-1} have been measured (Reiners & Basri 2008). This is an order of magnitude faster than the rotations of tidally-locked giant planets, and as a result coriolis forcing is more important in brown dwarf atmospheres. However, because of their greater surface gravities and photospheric pressures, the Rhines length and Rossby deformation radius scales (see contribution by Showman) in the upper atmospheres of brown dwarfs are roughly equivalent to those for hot Jupiters, of order the planetary/brown dwarf radius (assuming horizontal wind speeds comparable to the local sound speed, ~ 1 km s^{-1}; see Showman et al. 2008). As such, the small-scale banding and storm vorticities that characterize Jupiter's visible atmosphere are probably not common on either brown dwarfs or hot Jupiters, although detailed modeling of the former have yet to be reported.

In summary, while the current populations of brown dwarfs and (warm/hot) exoplanets may have photospheres with similar temperatures, significant differences in gas pressures and compositions may drive markedly dissimilar molecular chemistry. There is some overlap in temperature/pressure space between the densest/most massive exoplanets and the youngest/least-massive brown dwarfs, where meaningful comparisons in atmospheric properties and processes may be fruitfully made. The external forcing of a host star also results in exoplanetary atmospheric processes not seen in brown dwarfs: modified pressure-temperature profiles, inversion layers, photochemical production and thermal asymmetries that drive winds and jets. Yet at least in terms of flow dynamics, the atmospheres of both tidally-locked hot Jupiters and brown dwarfs should be weakly banded and have few small-scale vorticities in contrast to Jupiter.

3. Detailed Brown Dwarf Results Relevant to Exoplanet Studies

I conclude my contribution with two examples of low-temperature atmospheric processes studied in detail in brown dwarf studies but not yet sufficiently constrained in exoplanetary studies: condensate cloud formation and nonequilibrium chemistry.

3.1. Condensate Cloud Formation

Condensed species present in the photospheres of L dwarfs arise naturally from equilibrium chemistry, proceeding from the more refractory species such as mineral oxides and silicates (below 2500 K), to ionic salts and sulfides (below 1000 K), to "organic" condensates including as $H_2O[s]$ and $NH_3[s]$ (below 300 K; see review by Lodders & Fegley 2006). The presence of condensed species in L dwarfs has been inferred indirectly by their very red colors and muted molecular absorption features (e.g., Allard et al. 2001). Direct detection of silicate grain absorption has recently been made possible by the *Spitzer Space Telescope* (Cushing et al. 2006).

That these species reside in cloud structures in brown dwarf atmospheres has be inferred from other indicators: elemental depletion at high altitudes and the absence of condensates in T dwarf spectra. In the first case, the gravitational settling of condensed grains removes these species from the ambient gas, preventing further chemical reactions at higher (cooler) altitudes. For example, K I absorption is particularly strong in T dwarf spectra, despite the fact that K should have condensed out into silicate grains such as $KAlSi_3O_8[s]$ (othroclase). This reaction is inhibited by the depletion of Al and Si at deeper layers through the formation of, e.g., $CaTiO_3[s]$ (perovskite) and $Al_2O_3[s]$ (corundum; Burrows & Sharp 1999; Lodders & Fegley 2006); hence elemental K persists.[3] The absence of condensate cloud absorption in T dwarf spectra can be explained if condensates are vertically confined in cloud structures which ultimately sink below the visible photosphere (e.g., Ackerman & Marley 2001; Tsuji 2002; Cooper et al. 2003; Woitke & Helling 2004). As it turns out, the disappearance of condensate clouds at the transition between L and T dwarfs is quite abrupt, suggesting dynamic effects may be critical for cloud evolution (e.g., Burgasser et al. 2002; Liu et al. 2006; Burgasser 2007; Cushing et al. 2008).

The >500 L dwarfs observed to date reveal substantial diversity in cloud-sensitive features, including near-infrared colors (>1 mag scatter in $J-K$ for a given spectral subtype) and the strength of silicate grain absorption (e.g., Kirkpatrick et al. 2000; McLean et al. 2003; Burgasser et al. 2008). These variations have been simplistically interrupted as a range in cloud "thicknesses" (e.g. Cushing et al. 2008), although it is likely that other properties, such as grain size distribution, grain compositions (e.g., Helling et al. 2008) and cloud surface coverage also contribute. The properties of brown dwarf clouds are almost certainly tied to (interrelated) secondary parameters of age, surface gravity, metallicity, and rotation, as suggested by empirical trends (e.g., Faherty et al. 2009). However, current atmospheric models generally treat cloud properties as an independent model parameter, so source-to-source variations and temporal evolution can only be treated in a somewhat ad-hoc manner. Nevertheless, there has been substantial improvement in the fidelity and complexity of condensate cloud models to address the wealth of observational data, progress that can be ported to the still underconstrained problem of condensate clouds in hot exoplanetary atmospheres.

3.2. Nonequilibrium Chemistry and Atmospheric Dynamics

The contribution of Marley touches upon nonequilibrium chemistry in brown dwarf atmospheres in considerable detail, so I present only the major results here for completeness. Nonequilibrium chemistry refers to the nonequilibrium abundances of species that occur when the timescale for diffusive gas flow is shorter than the timescales governing the relevant chemical reactions. In brown dwarfs, the two reactions that are most affected by nonequilibrium chemistry convert $CO \rightarrow CH_4$ and $N_2 \rightarrow NH_3$. CO and N_2 have strong bonds and long chemical timescales at low temperatures, so they can appear in excess, and CH_4 and NH_3 in depletion, as a result of diffusive flows. Such abundance anomalies are indeed observed (Noll et al. 1997; Oppenheimer et al. 1998; Saumon et al. 2006), and indicate diffusivity constants of 1-100 m^2 s^{-1}, in excess of flows expected from convective instabilities (Saumon et al. 2007). It is likely that nonequilbrium chemistry is present in exoplanetary atmospheres as well, potentially giving rise to azimuthal abundance variations in sources with large day/night asymmetries, and hence modulation of phase-resolved spectroscopy in variance with equilibrium chemistry models. Such effects should be specifically sought for in future phase-resolved, direct spectroscopic studies of transiting exoplanets.

Acknowledgments. I thank A. Showman for insightful conversations on planetary atmospheric flow scales, K. Lodders and M. Marley for consultation on theoretical topics, and M. Cushing for providing Figure 1.

---

[3]Condensate depletion is also seen in the atmospheres of Jupiter and Saturn as traced by the presence of $GeH_4[s]$ (germane) over $SiH_4[s]$ (silane), despite the much greater elemental abundance of Si in a solar gas mixture (Fegley & Lodders 1994).

References

Ackerman, A. S., & Marley, M. S. 2001, ApJ, 556, 872
Allard, F., Hauschildt, P. H., Alexander, D. R., Tamanai, A., & Schweitzer, A. 2001, ApJ, 556, 357
Allard, N. F., Allard, F., Hauschildt, P. H., Kielkopf, J. F., & Machin, L. 2003, A&A, 411, L473
Allers, K. N., et al. 2007, ApJ, 657, 511
Bakos, G. Á., et al. 2007, ApJ, 670, 826
Barber, R. J., Tennyson, J., Harris, G. J., & Tolchenov, R. N. 2006, MNRAS, 368, 1087
Basri, G., & Brown, M. E. 2006, Annual Review of Earth and Planetary Sciences, 34, 193
Berger, E. 2006, ApJ, 648, 629
Burgasser, A. J. 2007, ApJ, 659, 655
Burgasser, A. J., Burrows, A., & Kirkpatrick, J. D. 2006, ApJ, 639, 1095
Burgasser, A. J., Cruz, K. L., & Kirkpatrick, J. D. 2007, ApJ, 657, 494
Burgasser, A. J., Looper, D. L., Kirkpatrick, J. D., Cruz, K. L., & Swift, B. J. 2008, ApJ, 674, 451
Burgasser, A. J., Marley, M. S., Ackerman, A. S., Saumon, D., Lodders, K., Dahn, C. C., Harris, H. C., & Kirkpatrick, J. D. 2002, ApJ, 571, L151
Burningham, B., et al. 2008, MNRAS, 391, 320
Burrows, A., Hubbard, W. B., Lunine, J. I., & Liebert, J. 2001, Reviews of Modern Physics, 73, 719
Burrows, A., Marley, M., Hubbard, W. B., Lunine, J. I., Guillot, T., Saumon, D., Freedman, R., Sudarsky, D., & Sharp, C. 1997, ApJ, 491, 856
Burrows, A., & Sharp, C. M. 1999, ApJ, 512, 843
Burrows, A., & Volobuyev, M. 2003, ApJ, 583, 985
Chabrier, G., & Baraffe, I. 2000, ARA&A, 38, 337
Chauvin, G., Lagrange, A.-M., Dumas, C., Zuckerman, B., Mouillet, D., Song, I., Beuzit, J.-L., & Lowrance, P. 2004, A&A, 425, L29
Chauvin, G., Lagrange, A.-M., Zuckerman, B., Dumas, C., Mouillet, D., Song, I., Beuzit, J.-L., Lowrance, P., & Bessell, M. S. 2005, A&A, 438, L29
Cooper, C. S., Sudarsky, D., Milsom, J. A., Lunine, J. I., & Burrows, A. 2003, ApJ, 586, 1320
Cushing, M. C., Marley, M. S., Saumon, D., Kelly, B. C., Vacca, W. D., Rayner, J. T., Freedman, R. S., Lodders, K., & Roellig, T. L. 2008, ApJ, 678, 1372
Cushing, M. C., et al. 2006, ApJ, 648, 614
Dahn, C. C., et al. 2002, AJ, 124, 1170
Deleuil, M., et al. 2008, A&A, 491, 889
Delorme, P., et al. 2008, A&A, 482, 961
Dupuy, T. J., Liu, M. C., & Ireland, M. J. 2008, ArXiv e-prints
Faherty, J. K., Burgasser, A. J., Cruz, K. L., Shara, M. M., Walter, F. M., & Gelino, C. R. 2009, AJ, 137, 1
Fegley, B. J., & Lodders, K. 1994, Icarus, 110, 117
Fortney, J. J. 2007, Ap&SS, 307, 279
Fortney, J. J., Lodders, K., Marley, M. S., & Freedman, R. S. 2008a, ApJ, 678, 1419
Fortney, J. J., Marley, M. S., Saumon, D., & Lodders, K. 2008b, ApJ, 683, 1104
Freedman, R. S., Marley, M. S., & Lodders, K. 2008, ApJS, 174, 504
Gelino, C. R., Marley, M. S., Holtzman, J. A., Ackerman, A. S., & Lodders, K. 2002, ApJ, 577, 433
Goldman, B. 2005, Astronomische Nachrichten, 326, 1059
Golimowski, D. A., et al. 2004, AJ, 127, 3516
Gonzalez, G. 1997, MNRAS, 285, 403
Griffith, C. A., & Yelle, R. V. 1999, ApJ, 519, L85
Helling, C., Dehn, M., Woitke, P., & Hauschildt, P. H. 2008, ApJ, 675, L105
Helling, C., & Woitke, P. 2006, A&A, 455, 325
Kalas, P., Graham, J. R., Chiang, E., Fitzgerald, M. P., Clampin, M., Kite, E. S., Stapelfeldt, K., Marois, C., & Krist, J. 2008, Science, 322, 1345

Kirkpatrick, J. D. 2005, ARA&A, 43, 195
Kirkpatrick, J. D., Barman, T. S., Burgasser, A. J., McGovern, M. R., McLean, I. S., Tinney, C. G., & Lowrance, P. J. 2006, ApJ, 639, 1120
Kirkpatrick, J. D., Dahn, C. C., Monet, D. G., Reid, I. N., Gizis, J. E., Liebert, J., & Burgasser, A. J. 2001, AJ, 121, 3235
Kirkpatrick, J. D., Reid, I. N., Liebert, J., Gizis, J. E., Burgasser, A. J., Monet, D. G., Dahn, C. C., Nelson, B., & Williams, R. J. 2000, AJ, 120, 447
Kunde, V. G., et al. 2004, Science, 305, 1582
Lagrange, A. ., et al. 2008, ArXiv e-prints
Leggett, S. K., Saumon, D., Albert, L., Cushing, M. C., Liu, M. C., Luhman, K. L., Marley, M. S., Kirkpatrick, J. D., Roellig, T. L., & Allers, K. N. 2008, ApJ, 682, 1256
Liebert, J., Kirkpatrick, J. D., Reid, I. N., & Fisher, M. D. 1999, ApJ, 519, 345
Linsky, J. L. 1969, ApJ, 156, 989
Liu, M. C., Leggett, S. K., Golimowski, D. A., Chiu, K., Fan, X., Geballe, T. R., Schneider, D. P., & Brinkmann, J. 2006, ApJ, 647, 1393
Lodders, K., & Fegley, Jr., B. 2006, Chemistry of Low Mass Substellar Objects (Astrophysics Update 2), 1–+
Luhman, K. L. 1999, ApJ, 525, 466
Luhman, K. L., Joergens, V., Lada, C., Muzerolle, J., Pascucci, I., & White, R. 2007, Protostars and Planets V, 443
Marois, C., Macintosh, B., Barman, T., Zuckerman, B., Song, I., Patience, J., Lafreniere, D., & Doyon, R. 2008, ArXiv e-prints
Marley, M. S., & Leggett, S. K. 2008, in The Future of Ultracool Dwarf Science with JWST, ArXiv e-prints 0803.1476
Martín, E. L., Delfosse, X., Basri, G., Goldman, B., Forveille, T., & Zapatero Osorio, M. R. 1999, AJ, 118, 2466
Mayor, M., & Queloz, D. 1995, Nat, 378, 355
McLean, I. S., McGovern, M. R., Burgasser, A. J., Kirkpatrick, J. D., Prato, L., & Kim, S. S. 2003, ApJ, 596, 561
Metchev, S. A., & Hillenbrand, L. A. 2006, ApJ, 651, 1166
Mohanty, S., Basri, G., Jayawardhana, R., Allard, F., Hauschildt, P., & Ardila, D. 2004a, ApJ, 609, 854
Mohanty, S., Basri, G., Shu, F., Allard, F., & Chabrier, G. 2002, ApJ, 571, 469
Mohanty, S., Jayawardhana, R., & Basri, G. 2004b, ApJ, 609, 885
Mohanty, S., Jayawardhana, R., Huélamo, N., & Mamajek, E. 2007, ApJ, 657, 1064
Nakajima, T., Oppenheimer, B. R., Kulkarni, S. R., Golimowski, D. A., Matthews, K., & Durrance, S. T. 1995, Nat, 378, 463
Noll, K. S., Geballe, T. R., & Marley, M. S. 1997, ApJ, 489, L87
Oppenheimer, B. R., Kulkarni, S. R., Matthews, K., & van Kerkwijk, M. H. 1998, ApJ, 502, 932
Oppenheimer, B. R., Kulkarni, S. R., & Stauffer, J. R. 2000, Protostars and Planets IV, 1313
Rayner, J. T., Cushing, M. C., & Vacca, W. D. 2009, in preparation
Reiners, A., & Basri, G. 2008, ApJ, 684, 1390
Saumon, D., Bergeron, P., Lunine, J. I., Hubbard, W. B., & Burrows, A. 1994, ApJ, 424, 333
Saumon, D., Marley, M. S., Cushing, M. C., Leggett, S. K., Roellig, T. L., Lodders, K., & Freedman, R. S. 2006, ApJ, 647, 552
Saumon, D., et al. 2007, ApJ, 656, 552
Schilbach, E., Roeser, S., & Scholz, R. . 2008, ArXiv e-prints
Showman, A. P., Menou, K., & Cho, J. Y.-K. 2008, in Astronomical Society of the Pacific Conference Series, Vol. 398, Astronomical Society of the Pacific Conference Series, ed. D. Fischer, F. A. Rasio, S. E. Thorsett, & A. Wolszczan, 419
Smith, V. V., Tsuji, T., Hinkle, K. H., Cunha, K., Blum, R. D., Valenti, J. A., Ridgway, S. T., Joyce, R. R., & Bernath, P. 2003, ApJ, 599, L107
Sozzetti, A., Torres, G., Charbonneau, D., Latham, D. W., Holman, M. J., Winn, J. N., Laird, J. B., & O'Donovan, F. T. 2007, ApJ, 664, 1190

Tinetti, G., Liang, M.-C., Vidal-Madjar, A., Ehrenreich, D., Lecavelier des Etangs, A., & Yung, Y. L. 2007, ApJ, 654, L99
Torres, G., Winn, J. N., & Holman, M. J. 2008, ApJ, 677, 1324
Tsuji, T. 2002, ApJ, 575, 264
Vrba, F. J., et al. 2004, AJ, 127, 2948
Whitworth, A., Bate, M. R., Nordlund, Å., Reipurth, B., & Zinnecker, H. 2007, Protostars and Planets V, 459
Wilson, J. C., Kirkpatrick, J. D., Gizis, J. E., Skrutskie, M. F., Monet, D. G., & Houck, J. R. 2001, AJ, 122, 1989
Woitke, P., & Helling, C. 2004, A&A, 414, 335
Zapatero Osorio, M. R., Lane, B. F., Pavlenko, Y., Martín, E. L., Britton, M., & Kulkarni, S. R. 2004, ApJ, 615, 958

Radiation Hydrodynamics Simulations of Dust Clouds in the Atmospheres of Substellar Objects

Bernd Freytag and France Allard

Centre de Recherche Astrophysique de Lyon, UMR 5574: CNRS, Université de Lyon, École Normale Supérieure de Lyon, 46 allée d'Italie, F-69364 Lyon Cedex 07, France

Derek Homeier

Institut für Astrophysik Göttingen, Georg-August-Universität, Friedrich-Hund-Platz 1, D-37077 Göttingen, Germany

Hans-Günter Ludwig

Observatoire de Paris-Meudon, GEPI-CIFIST, 92195 Meudon, France

Matthias Steffen

Astrophysikalisches Institut Potsdam, An der Sternwarte 16, D-14482 Potsdam, Germany

Abstract. The temperature structure and the motions in the atmospheres of cool stars are affected by the underlying convection zone. The radiation hydrodynamics code CO5BOLD has been developed to simulate (small patches of the) convective surface layers of these stars. Updated opacity tables based on PHOENIX data and a description for the formation, destruction, advective transport, and settling of dust have made the code fit to handle the conditions in brown dwarf atmospheres. Currently, objects from 8500 K down to about 900 K have been simulated. Recently, incident radiation has been included, allowing simulations with conditions found on hot planets. In non-irradiated brown dwarf models we encounter mixing by gravity waves and in the cooler models convection within the clouds. The qualitative effects of incident radiation are surprisingly small, as long as the effective temperature of the object stays well below the dust condensation temperature. Beyond that point, there are no layers where dust could form, anymore.

1. Introduction

Temperatures in the atmospheres of late type M dwarf stars and brown dwarfs are so low that dust particles can form ($T_{gas} < 1800K$). These grains should sink under the influence of gravity into deeper layers and vanish from the atmosphere, clearing it from condensable material. However, observed spectra can only be reproduced by models accounting for dust formation and its resulting greenhouse effect in the visible layers (Tsuji et al. 1996; Leggett et al. 1998). The approaches to model dust within classical 1D hydrostatic stellar atmosphere models presented in Helling et al. (2008) differ considerably, and all rely on not well justified assumptions about the extent of the cloud layers or the amount of mixing.

On candidate to cause mixing is surface convection that influences the stratification in various ways: Convective energy transport leads to a change in the mean temperature structure but also to spatial and temporal fluctuations in the temperature field, with implications for the formation of molecules (see e.g. Wedemeyer-Böhm et al. 2005). Velocity fields affect shapes of spectral lines, are able to mix material, and can transport wave energy. Moreover, the correct description of the surface convection zone is necessary for detailed models of the interior of stellar or sub-stellar objects.

Time-dependent radiation hydrodynamics (RHD) models can describe self-consistently the mixing of material beyond the classical boundaries of a convection zone, as demonstrated for instance for main-sequence A-type stars (Freytag et al. 1996) or for M dwarfs (Ludwig et al. 2002, 2006). The aim of the current project is to extend the latter simulations into the regime of brown dwarfs and hot planets, where dust clouds have a strong influence onto the photospheric temperature structure.

2. Simulations with CO5BOLD

2.1. Numerical Radiation Hydrodynamics

We used the radiation-hydrodynamics code CO5BOLD in the local setup (see Freytag et al. 2002; Wedemeyer et al. 2004; Freytag et al. 2004)
to calculate time-dependent 2D models of the atmospheres of sub-stellar objects, including the very top layers of the convection zone. The restriction to two dimensions, so far, arises from the need to cover the extremely long settling and mixing time scales of dust and monomers. The code solves the coupled equations of compressible hydrodynamics and non-local radiation transport on a Cartesian grid with a time-explicit scheme. The tabulated equation of state account for the ionization of hydrogen and helium, and the formation of H_2 molecules.

The 1D hydrodynamics fluxes are computed with an approximate Riemann solver of Roe-type and combined unsplit, i.e. the fluxes in vertical and horizontal direction are computed from the same state (and not after each other) and their contributions are added, because the conditions in the cool objects are almost incompressible.

Usually, we use open top and bottom boundary conditions for local models that comprise the upper part of a deep convection zone. However, closed boundaries keep the amount of dust constant within the computational domain and are employed for all the brown dwarf models, i.e., the lower boundary is closed although the stellar convection zone should extend down to the center of the star. To keep the entropy close to a prescribed value, the internal energy in a few grid layers (10 km high) at the bottom of the model is adjusted. This mechanism acts as an energy source and replenishes the radiative energy losses through the top of the model. This parameter (the value of the entropy plateau s_{in} in the deep convective layers) controles the effective temperature and is taken from the start model. In addition, a drag force damps downdrafts in these layers.

The top boundary is closed, too, partly to keep material inside. It has damping zone of about 8 grid points where a strong drag force is applied. Damping just at an open boundary appeared to be not sufficient to keep under control gravity waves with moderate Mach number (with peak values close to 1) but large vertical gradients due to the occurrence of waves with small vertical wavelength.

Incident radiation along the vertical ray is described by two parameters: the effective temperature of a black body $T_{\text{eff,inci}}$ and a dilution factor $(R_{\text{star}}/d_{\text{star-planet}})^2$.

2.2. Dust Model

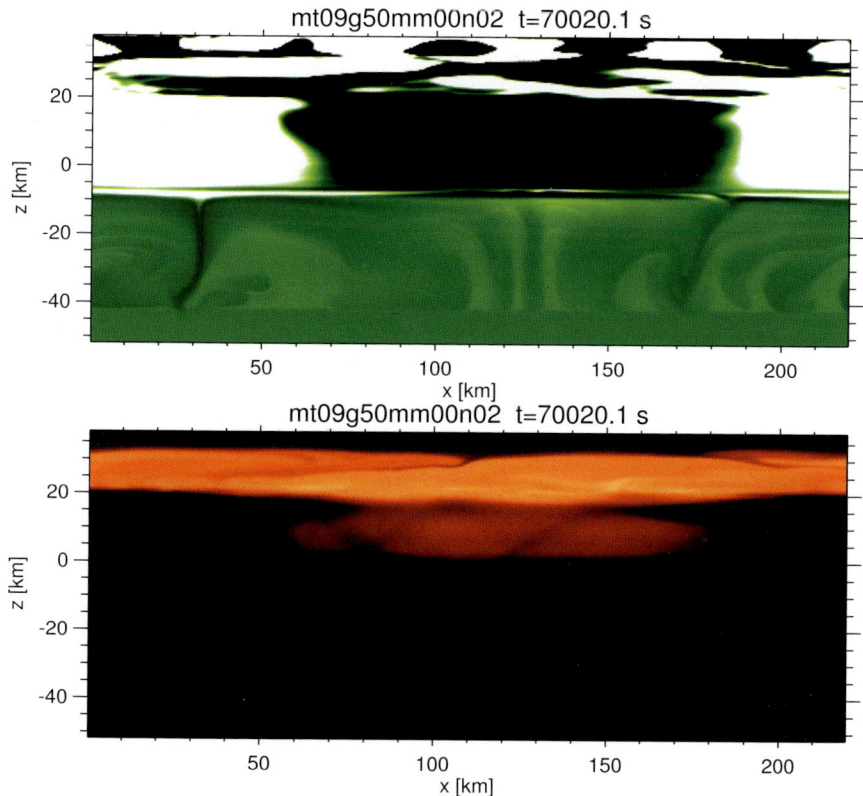

Figure 1. Entropy fluctuations (top panel; light shades (bright color) indicates material hotter than the surroundings) and dust concentration (bottom panel; the bright band at the top is the dust), without irradiation. T_{eff}=900 K, log$g$=5.

To account for the presence of dust we added terms in the modules for hydrodynamics, radiation transport, source terms, and in the handling of boundary conditions.

It is impossible to account for all microphysical processes that might play a role for dust formation in current time-dependent multi-dimensional simulations. Instead, we chose a treatment of dust that includes only the most basic – and hopefully the most important – ingredients.

We concentrate on Forsterite grains (Mg_2SiO_4, 3.3 g/cm^3) that are relatively abundant and give the largest contribution to the total dust opacities. The dust scheme is based on a simplified version of the dust model used in Höfner et al. (2003): one density field (instead of four as in Höfner et al. 2003 and Freytag & Höfner 2008) describes the

128   Freytag et al.

amount of dust, the other the amount of "monomers". The ratio of dust plus monomers density over gas density is allowed to change, in contrast to the dust description by Höfner et al. (2003).

Instead of modelling the nucleation and the detailed evolution of the number of grains we assume a constant fraction of seeds in the total amount of dust and monomers. Would all material in a grid cell be condensed into dust, the grains would have the maximum radius $r_{d,max}$, which is an external parameter and is set typically to $1\,\mu$m. The radius r_d of dust grains for given dust mass density ρ_d and monomer mass density ρ_m is computed from $r_d = r_{d,max}\,[\rho_d/(\rho_d + \rho_m)]^{1/3}$. Condensation and evaporation are modelled as in Höfner et al. (2003), with parameters and saturation vapor curve adapted to Forsterite.

In the hydrodynamics module, monomers and dust densities are advected together with the gas density. However, a settling speed according to Rossow (1978) is added to the vertical advection velocity of the dust grains, assuming instantaneous equilibrium between gravitational and viscous force that act onto the grains.

For RHD models of cool stars we usually assume a complete mixture of the atmosphere and the same element composition everywhere in the model. CO5BOLD can deal with the effects of ionization but not with a variation of element composition with space and time. Therefore, the depletion of elements is completely ignored in the equation of state: anyway, the formation of molecules has only a minor effect on e.g. the heat capacity as long as hydrogen exists in the form of H_2. However, molecules play a major role for the opacity. And the formation of molecules depends on the abundance and depletion of elements. To take this partially into account we derive the CO5BOLD opacity table from a data cube with $\kappa(T, P, \nu)$ generated with the PHOENIX atmosphere code (Ferguson et al. 2005), since gas phase opacity calculations are too prohibitive for a dynamical account in hydro simulations. For this application we find the limiting *Cond* case a priori most suitable, i.e. dust condensation in equilibrium with the gas phase while ignoring dust opacities so as to reproduce an efficient sedimentation case. The resulting gas phase opacities are depleted from refractory element contributions in the proportion predicted by chemical equilibrium in layers were conditions favor dust formation, and not depleted in inner atmospheric layers where gas is too hot for dust formation. This choice allows for an as adequate as possible account of gas phase opacities in the innermost atmospheric layers where dust does not form, as well as in the dust forming layers. Dust opacities are included dynamically. The detailed gas opacities are averaged into 5 bins to account for non-grey effects. In contrast to the sophisticated treatment of the gas opacities, we use a simple formula for the dust opacities, assuming that the large particle limit is valid for all grain sizes and treating scattering as true absorption.

3. Results of the Simulations

3.1. Temperature Sequence

We present a temperature sequence of non-irradiated models reaching from cool M dwarfs to cool brown dwarfs (T_{eff}=2800 K to 900 K; log$g$=5) demonstrating the contributions of different mixing processes.

The top panel in Fig. 1 shows the typical granulation pattern with cool downdrafts even narrower than in solar granulation (Ludwig et al. 2002, 2006) in a warmer

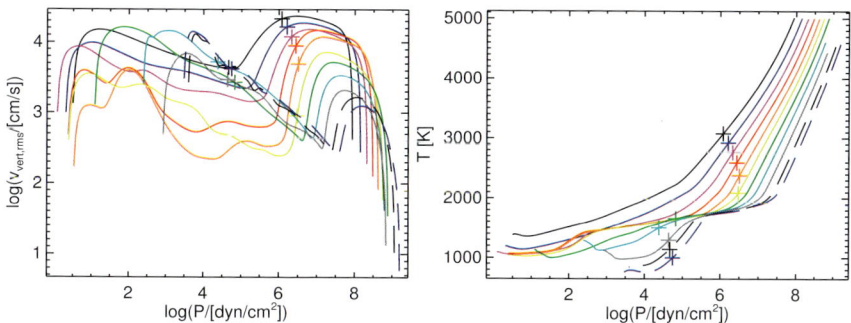

Figure 2. Logarithm of rms-value of vertical velocity (left) and mean temperature (right) over logarithm of pressure for models with effective temperatures (from top to bottom at the right part of both plots) of 2800 K, 2600 K, 2400 K, 2200 K, 2000 K, 1800 K, 1600 K, 1400 K, 1200 K, 1000 K, 900 K. Plus signs indicate the points where the Rosseland optical depths reaches unity.

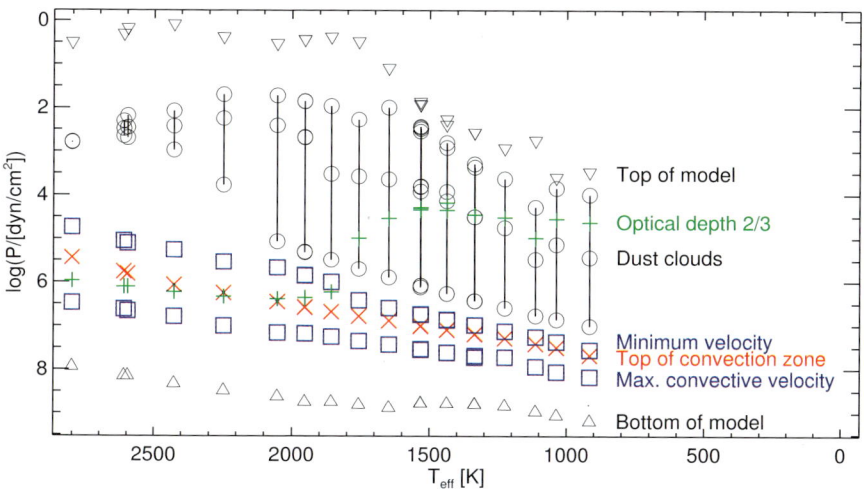

Figure 3. Logarithm of pressure over effective temperature for various points of interest: black triangles: top and bottom of each model, red crosses: top of convectively unstable layers, lower set of blue squares: point withy maximum convective velocity $v_{\text{vert,rms}}$, black circles: location of cloud layers, green plus signs: layers with $\tau_{\text{Rosseland}}=2/3$.

environment. There is a thin transition layer to the stable atmosphere with strong inhomogeneities induced by *gravity waves*. Figure 1(bottom) indicates the location of the actual dust clouds and shows their spatial variations due to gravity waves and convection within the clouds.

The bump on the plot of the rms-value of the vertical velocity over the logarithm of pressure in Fig. 2 (left) is due to convective motions. The wiggly increase with decreasing pressure (to the left) is mostly due to gravity waves. While the amplitude of the convection in the deeper layers decreases monotonically with effective temperature, the wave amplitudes show a more complex temperature dependence.

The plot of the temperature over the logarithm of pressure in Fig. 2 (right) shows the effect of the dust onto the stratification for models of 2400 K and below: within the clouds the temperature drops rapidly, while below the cloud deck the greenhouse effect causes a temperature plateau with a fairly shallow slope.

Figure 3 shows the position of the clouds relative to the convection zone for the entire temperature sequences covered by the models. At higher effective temperatures one can just see through the thin cloud layers right into the top of the convection zone. A intermediate temperatures the clouds become thicker and opaque. At lower effective temperatures most of the dust sits below the visible layers which are less affected by it.

3.2. Sequences with Different Amounts of Irradiation

We varied the incident radiation for models with two different internal effective temperatures ($T_{\rm eff}$=1200 K and 900 K), resulting in a models with various total effective temperature $T_{\rm eff,t}$, as indicated in the plots.

We started with small amounts of irradiation being afraid that a significant flux entering the model at the top would somehow impede the convection and alter the entire structure. However, small amount of incident flux have a small effect (compare Figs. 1 and 4) essentially restricted to the upper atmospheric layers only. As seen in the top right panels in Fig. 6 and Fig. 7 the temperature rises slightly above and at the top of the cloud layers until the effective temperature is sufficient to emit the flux coming from below as well as the irradiated flux. There is only little effect onto the velocities (top left panels in Figs. 6 and 7) and the dust concentration decreases only slightly (bottom left panels), as also observed with Phoenix Dusty models (Barman et al. 2001). The decrease in dust leads to a decrease in the monomer depletion (bottom right panels).

In contrast, the irradiation in the CoRoT-3b case (model with $T_{\rm eff,t}$=1735 K in Fig. 7) is sufficient to raise the temperature to levels where essentially no (Forsterite) dust can form and the corresponding monomers show little depletion. We find however a dust haze forming close to the top. The velocity amplitude shows little change. The remaining atmospheric motions are purely due to gravity waves (cf. Fig. 5).

4. Conclusions

The presented 2D radiation hydrodynamical atmosphere models of substellar objects show that overshoot velocities decline roughly exponentially with distance from the convectively unstable regions, as found previously in e.g. A-type and M-type dwarfs. However, the velocities drop so steeply that they are only important close to the convection zone. Instead, *gravity waves* dominate the mixing of the upper atmospheric layers with amplitudes even growing with height, modulating dust concentration and

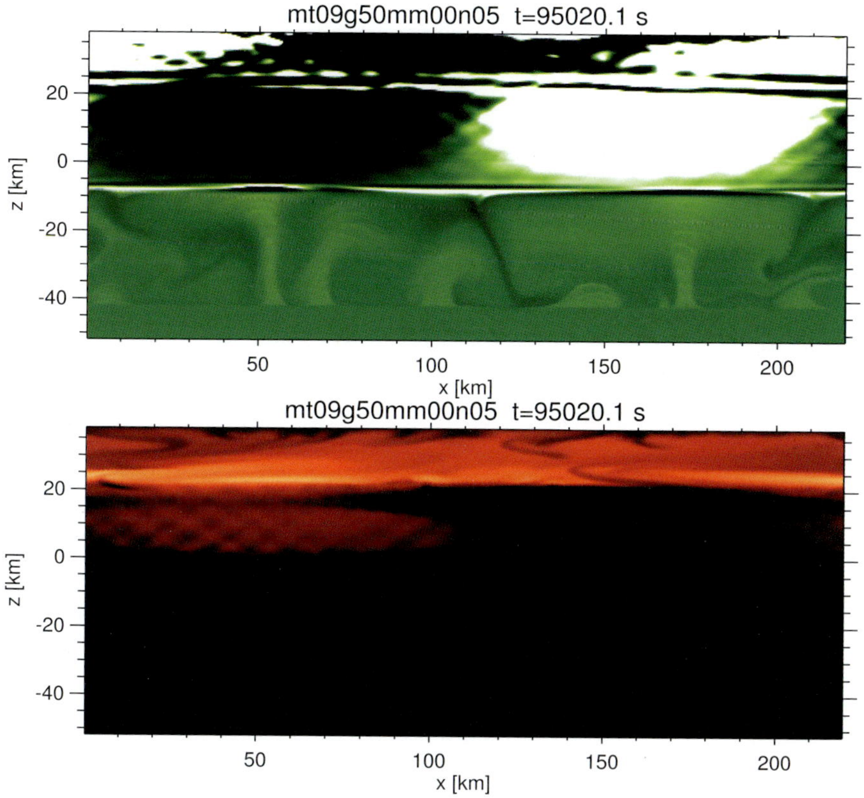

RHD Simulations of Dust Clouds 131

Figure 4. Entropy fluctuations and dust concentration for model with small amounts of irradiation: $\log g = 5$, $T_{\rm eff} = 900$ K, $T_{\rm eff,t} = 1276$ K.

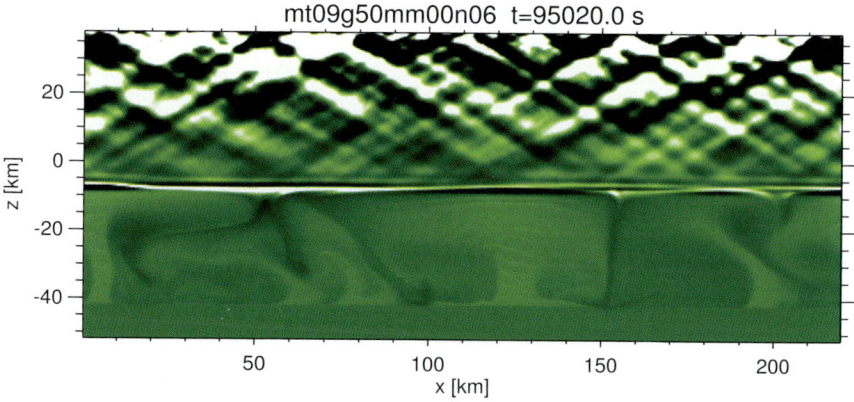

Figure 5. Entropy fluctuations for model with strong irradiation. $\log g = 5$, $T_{\rm eff} = 900$ K, $T_{\rm eff,t} = 1735$ K.

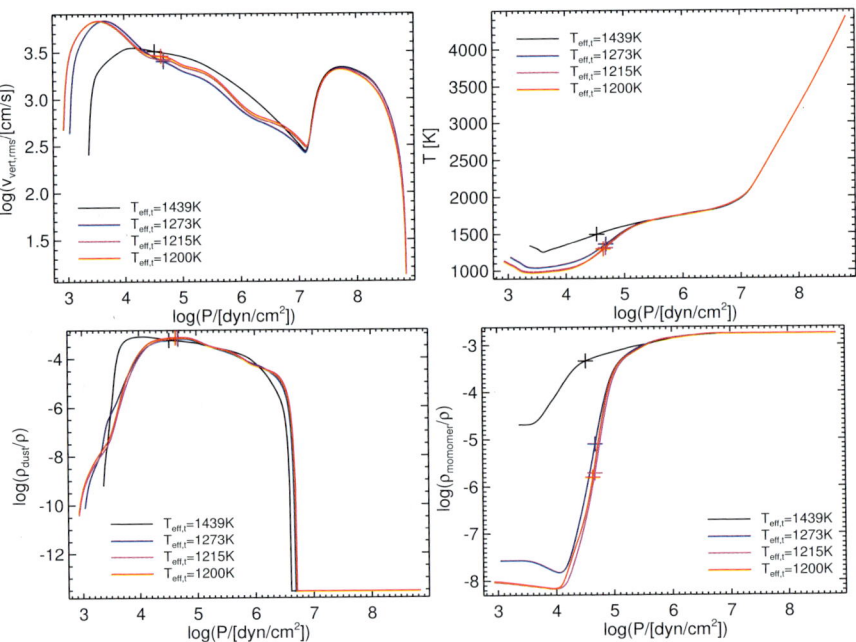

Figure 6. Logarithm of rms-value of vertical velocity (top left), mean temperature (top right), dust concentration (bottom left), and monomer concentration (bottom right) over logarithm of pressure for models with different amount of irradiation. The internal effective temperature is of about 1200 K. Gravity log$g$=5.

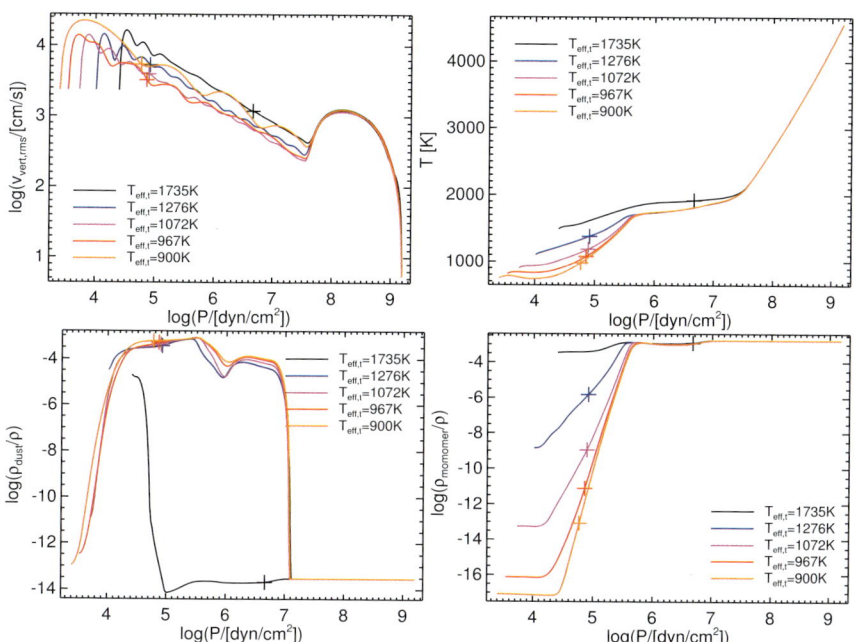

Figure 7. Logarithm of rms-value of vertical velocity (top left), mean temperature (top right), dust concentration (bottom left), and monomer concentration (bottom right) over logarithm of pressure for models with different amount of irradiation. The internal effective temperature is of about 900 K. Gravity $\log g=5$.

cloud thickness. At effective temperatures below about 2000 K, dust layers become thick enough to cause *convection within the clouds*. The induced combined mixing is sufficient to balance the settling of dust grains.

Models with higher effective temperatures show a high-altitude haze of optically thin clouds. At lower effective temperatures thick and dense clouds exist but mostly below the visible layers, that are essentially depleted of the material that went into the dust. In between, dust is an important opacity source in the atmosphere.

Incident radiation mainly affects the temperature of the uppermost atmospheric layers. Small amounts of irradiation (corresponding to orbital distances beyond about 0.05 AU for a solar type star host) have no qualitative effect onto the cloud dynamics, causing only a small rise of the temperatures at the top and above the clouds, which leads to some decrease of the dust concentration.

Nevertheless, when irradiation heats up the atmosphere too much, as is the case for CoRoT-3b at 0.055 AU from an F star host, dust cloud formation and with it cloud convection is suppressed, except perhaps in the planetary limb i.e. a dust haze in the uppermost layers. But gravity waves induced by the lower convection zone remain.

References

Barman, T. S., Hauschildt, P. H., & Allard, F. 2001, ApJ, 556, 885
Ferguson, J. W., Alexander, D. R., Allard, F., et al. 2005, ApJ, 623, 585
Freytag, B. & Höfner, S. 2008, A&A, 483, 571
Freytag, B., Ludwig, H.-G., & Steffen, M. 1996, A&A, 313, 497
Freytag, B., Steffen, M., & Dorch, B. 2002, Astronomische Nachrichten, 323, 213
Freytag, B., Steffen, M., Wedemeyer-Böhm, S., & Ludwig, H.-G. 2004, CO5BOLD User Manual, http://www.astro.uu.se/~bf/co5bold_main.html
Helling, C., Ackerman, A., Allard, F., et al. 2008, MNRAS, 1310
Höfner, S., Gautschy-Loidl, R., Aringer, B., & Jørgensen, U. G. 2003, A&A, 399, 589
Leggett, S. K., Allard, F., & Hauschildt, P. H. 1998, ApJ, 509, 836
Ludwig, H.-G., Allard, F., & Hauschildt, P. H. 2002, A&A, 395, 99
Ludwig, H.-G., Allard, F., & Hauschildt, P. H. 2006, A&A, 459, 599
Rossow, W. B. 1978, Icarus, 36, 1
Tsuji, T., Ohnaka, K., & Aoki, W. 1996, A&A, 305, L1
Wedemeyer, S., Freytag, B., Steffen, M., Ludwig, H.-G., & Holweger, H. 2004, A&A, 414, 1121
Wedemeyer-Böhm, S., Kamp, I., Bruls, J., & Freytag, B. 2005, A&A, 438, 1043

J-P. Maillard: session chairman.

Part VI

Terrestrial Exoplanets: Modelling, Habitability, and Detection of Biosignatures (Chair: Ignasi Ribas)

The Loss of Nitrogen-rich Atmospheres from Earth-like Exoplanets within M-star Habitable Zones

H. Lammer, H. I. M. Lichtenegger and M. L. Khodachenko

Space Research Institute, Austrian Academy of Sciences, Schmiedlstr. 6, A-8042 Graz, Austria

Yu. N. Kulikov

Polar Geophysical Institute (PGI), Russian Academy of Sciences, Khalturina Str. 15, Murmansk, 183010, Russian Federation

J.-M. Grießmeier

ASTRON, P.O. Box 2, 7990 AA Dwingeloo, The Netherlands

Abstract. After the first discovery of massive Earth-like exoplanets around M-type dwarf stars, the search for exoplanets which resemble more an Earth analogue continues. The discoveries of super-Earth planets pose questions on habitability and the possible origin of life on such planets. Future exoplanet space projects designed to characterize the atmospheres of terrestrial exoplanets will also search for atmospheric species which are considered as bio-markers (e.g. O_3, H_2O, CH_4, etc.). By using the Earth with its atmosphere as a proxy and in agreement with the classical habitable zone concept, one should expect that Earth-like exoplanets suitable for life as we know it should have a nitrogen atmosphere and a very low CO_2 content. Whether a water bearing terrestrial planet within its habitable zone can evolve into a habitable world similar than the Earth, depends on the capability of its water-inventory and atmosphere to survive the period of high radiation of the young and/or active host star. Depending on their size and mass, lower mass stars remain at high X-ray and EUV (XUV) activity levels for hundreds of Ma's to Ga's. XUV flux values which are 10 or 20 times higher than that of the present Sun can heat the thermosphere and expand the exobase of N_2-rich Earth-like exoplanets to altitudes well above their expected magnetopause distances. This results in magnetically non-protected upper atmospheres and high non-thermal escape rates. We studied this plasma induced N^+ ion pick up escape and applied a numerical test-particle stellar wind plasma - exosphere interaction model. Our results indicate that Earth-analogue exoplanets with atmosphere compositions similar to that of present Earth will lose their nitrogen inventories if they are exposed over a sufficient period of time to XUV fluxes ≥ 10 times that of the present Sun. Because most M-type stars are active in XUV radiation we suggest that these planets will undergo a different atmospheric evolution than the Earth so that life as we know it may not evolve on their surfaces.

1. Introduction

Atmospheric escape from a planet during the active period of its host star is an important factor which influences the evolution of atmospheres and planetary water inventories.

Based on the energy source, escape processes can be separated into thermal escape and non-thermal escape.

Kulikov et al. (2006) and Lammer et al. (2007) found that high XUV radiation of the active young Sun or active dwarf stars results in considerable expansion of CO_2-rich thermospheres and exospheres of terrestrial planets. In case the upper atmosphere expands beyond a protecting magnetosphere (magnetopause), the atmosphere of an exoplanet which is exposed to high XUV fluxes and steller plasma flows can be in a real danger of being stripped of its whole gaseous envelope even if the planet orbits its parent M star within the habitable zone (HZ). However, as shown in Lammer et al. (2007), a high CO_2 atmospheric mixing ratio will result in enhanced IR cooling in the thermosphere and inhibits its expansion and therefore leads to reduced non-thermal atmospheric erosion due to dense stellar winds or Coronal Mass Ejections (CME's).

Recently, Tian et al. (2008) developed a multi-component hydrodynamic thermosphere model where they could self consistently study the present time Earth thermosphere under extreme solar XUV conditions. Their model was validated against observations (MSIS-00) and models (e.g., Roble et al. 1987; Smithtro and Sojka 2005a; Smithtro and Sojka 2005b) of the Earth's present N_2-rich thermosphere. Tian et al. (2008) found that a thermosphere of an Earth-size and mass planet with the present time Earth gas composition can experience a hydrodynamic expansion, so that hydrostatic equilibrium can no longer be maintained. Their results indicate that hydrodynamic flow of the bulk atmosphere and the associated adiabatic cooling cannot be ignored when one studies the response of thermospheres to high stellar XUV radiation.

Furthermore, these authors show that the fast variation of the thermospheric bulk motion velocities under different XUV induced temperatures suggest that the adiabatic cooling effect could keep the exobase temperature $\leq$ 1000 K if light gases such as atomic hydrogen are the dominant species in an thermosphere.

From the model results of Tian et al. (2008) one can conclude that the upper atmosphere of an Earth-mass planet having the same atmospheric composition than present Earth and exposed to extreme XUV fluxes $\geq$ 10 times that of the present Sun would start to rapidly expand if thermospheric temperatures exceeded values of about 7000-8000 K, when atomic O and N dominate the upper thermosphere. Due to the heating, the atmosphere may even expand beyond the magnetopause stand-off distance as illustrated in Fig. 1. However, this process is limited due to the outflow of the dominant species which in turn results in adiabating cooling of the atmosphere.

The finding of Tian et al. (2008) is very relevant for Earth-like exoplanets within orbits of M star HZ's, because these low mass stars are active over long time periods in the XUV wavelength range due to strong magnetic fields, which are generated by their fully or partially convective interiors (e.g. Ayres 1997; Audard et al. 2004; Ribas et al. 2005; Scalo et al. 2007; and references therein).

Low mass dwarf stars are more numerous in the solar neighbourhood than solar-like G-stars, therefore, they are also considered as interesting candidates for planet finding projects and missions. Furthermore, due to a smaller size of these stars they are less luminous which results in a closer orbital distance of their HZ's (e.g., Khodachenko et al. 2007). Although one can not generalize dwarf stars because their mass-range spread is wider compared with Sun-like spectral type stars, it is certain that terrestrial-type exoplanets within HZs of M-type stars are more influenced by the stellar radiation and particle environment than a planet in a HZ of a solar-like star.

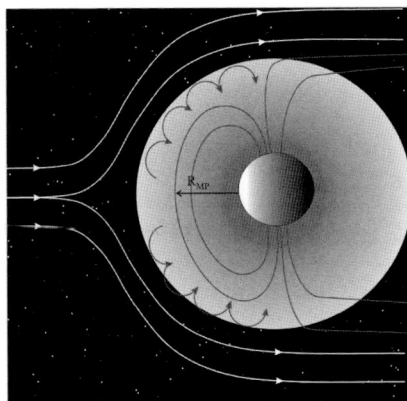

Figure 1. Illustration showing the expected stellar wind – atmosphere interaction in a case where the upper atmosphere expands above a compressed magnetosphere. Neutral species above the magnetopause can be ionized and picked up by the stellar wind plasma flow.

The focus of this study is to investigate if nitrogen-rich Earth-like exoplanets within M star HZs can keep their atmospheres or if they suffer strong stellar plasma wind induced atmospheric erosion. In Sect. 2 we briefly explain the input parameters and the numerical test particle model used for the ion pick up estimation and in Sect. 3 we discuss our results and the consequences for habitability.

2. N^+ Pick Up Loss Modeling

For studying the stellar plasma induced N^+ pick up escape rates we use for our model input the neutral atmosphere profiles, which were modeled by Tian et al. (2008). Their model treats the whole thermosphere as a single fluid with a varying mean molecular mass. For this moving atmosphere, the model iteratively solves the momentum and energy equations and includes neutral and ion species. Time independent diffusion equations are solved for the following long-lived species O, O_2, N_2, He, H, H_2, CO_2, CO, $N(^4S)$, NO, H_2O, O^+, N^+, and H^+. Chemical equilibrium is applied to 18 short-lived species (Tian et al. 2008). They assume quasi-charge neutrality and the model includes 154 chemical reactions, collected from thermospheric models of the present Earth (Roble 1995; Smithtro and Sojka 2005a), Venus and Mars (Nagy et al. 1983; Barth et al. 1992; Fox and Delgano 1979), and extrasolar planets (Yelle 2004). The lower boundary is at the mesopause at about 95 km and the upper boundary is the exobase, which moves to higher distances with increasing stellar XUV flux. Hydrodynamics is valid in the collision dominated thermosphere below the exobase. Above the exobase the particles are treated kinetically.

In agreement with Smithtro and Sojka (2005b), the results of Tian et al. (2008) indicate that under extreme XUV conditions atomic nitrogen becomes a competing species with atomic oxygen at the exobase level. Table 1 shows the corresponding exobase distances, temperatures and atomic nitrogen densities for 7 times, 10 times and 20 times that of the present Sun at a distance of 1 AU (within the HZ).

I_{EUV}/I_{Sun}	z_{exo}	T_{exo} [K]	n_N [cm^{-3}]
7	2.5 r_{Earth}	8000	10^5
10	4.8 r_{Earth}	5700	4.8×10^4
20	12.7 r_{Earth}	2500	1.6×10^4

Table 1. Exobase distance, temperature and nitrogen number density as a function of stellar EUV flux normalized to 1 AU for present Sun (Tian et al. 2008).

For estimating the number density in the collisionless exosphere above the exobase levels given in Table 1, we use Chamberlain's approach (e.g., Chamberlain 1963; Lammer et al. 2007). This approach is based on Liouville's equation, where the exospheric nitrogen number density as a function of the planetocentric distance can be written as the product of the barometric density and the sum of the partition functions that correspond to escaping, ballistic, and satellite particle trajectories (Chamberlain 1963).

Khodachenko et al. (2007) studied the influence of dense plasma flows (stellar winds and CME activity) on the expected planetary and magnetospheric environments of Earth-like exoplanets in orbital locations within close-in HZs and concluded that dense plasma flows could play a critical role in the definition of the habitability criterion for the evolution of terrestrial exoplanets orbiting dwarf stars.

Due to tidal locking and, hence, slow planetary rotation, Earth-like exoplanets within M star HZs are expected to have weaker magnetic dynamos compared to the faster rotating Earth at 1 AU (Grießmeier et al. 2005; Khodachenko et al. 2007). The dynamic pressure of denser and faster plasma winds acts against the planetary magnetic pressure, so that the magnetosphere is compressed and the resulting atmosphere protecting area is smaller. To investigate if smaller or more compressed magnetospheres of nitrogen-rich terrestrial exoplanets will result in little or no magnetospheric protection against the stellar plasma flow, we estimate the magnetopause stand off distance by assuming the plasma parameters shown in Table 2.

To make optimistic loss estimations, we assume a planet with the size and mass of the Earth and a faster rotation period of about 13 hours. The faster rotation assumed in our study will enhance the strength of the intrinsic magnetic moment of our Exo-Earth to about $1.5 \times \mathfrak{J}_{Earth}$, with $\mathfrak{J}_{Earth} = 6.6845 \times 10^{22}$ Am^2 being the magnetic moment of the Earth.

I_{XUV}/I_{Sun}	v_{sw} [km/s]	n_{sw} [cm^{-3}]	r_{mag}	r_{exo}
7	780	90	$6.0 r_{Earth}$	$2.5 r_{Earth}$
10	900	200	$5.2 r_{Earth}$	$4.8 r_{Earth}$
20	1010	270	$4.7 r_{Earth}$	$12.7 r_{Earth}$

Table 2. Stellar XUV flux normalized to present day Sun, stellar plasma parameters, magnetopause and exobase distances.

Because HZs of M-stars are closer to the star compared with G-type stars, one can expect higher plasma densities(winds, CME's) (Khodachenko et al. 2007). The values shown in Table 2 correspond to orbital distances between 0.2 - 0.4 AU. M star planets within closer HZ's are exposed to denser stellar plasma flows. One can see from Table 2 that due to the higher plasma pressure the magnetopause boundaries of our test planets should be closer to the planetary surfaces than at Earth even if we assume a 1.5 times stronger magnetic moment. At Earth the average solar activity induced magnetopause stand-off distance in 1 AU is located at about 10 Earth-radii and corresponds to average solar wind plasma densities of about 8 cm^{-3} and velocities of about 450 km/s.

One can see from Table 2 that only the 7 XUV case in connection with a stellar wind plasma density which is about 10 times that of the present Earth builds up a pressure balance between the stellar wind plasma and the magnetosphere at a distance which is $\geq 2.5 r_{Earth}$ above the exobase. Because we assume a denser stellar plasma flow for the 10 XUV case the dynamic pressure compresses the magnetosphere to a distance which is comparable to the expanded exobase level. For the 20 XUV case and the assumed stellar plasma flow parameters the magnetic moment is to weak for balancing the stellar wind dynamic pressure above or close to the exobase level. This result indicates that magnetospheres of nitrogen-rich Earth-like exoplanets which are exposed to XUV fluxes $\geq$ 10 XUV are not protecting the expanded upper atmosphere against stellar plasma erosion. In such a case one can expect that an ionopause will build up like at present day Venus or Mars around the exobase level (see also Yamauchi and Wahlund 2007).

We consider now the production rate of planetary N^+ ions as the sum of the rates of photo-ionization,

$$h\nu + N \rightarrow N^+ + e^-, \qquad (1)$$

electron impact ionization,

$$e^- + N \rightarrow N^+ + 2e^-, \qquad (2)$$

and charge exchange

$$p_{sw}^+ + N \rightarrow N^+ + H^{ENA}, \qquad (3)$$

between the atomic nitrogen exosphere above the planetary obstacle (magnetopause or ionopause) and the stellar plasma flow.

The simulation of the particle escape is initialized by dividing the space around the planetary obstacle into a number of volume elements. Production rates of planetary N^+ ions are then obtained by calculating the ion production along streamlines around the planetary obstacle (shown in Fig. 2). Upon assuming that all newly created N^+ ions will be picked up by the stellar wind and carried away from the planet, the loss rate in each volume element can be determined. Finally, by summing over all elements the total N^+ pick up escape rate is obtained (e.g., Lammer et al. 2003; Erkaev et al. 2005; Lammer et al. 2007).

3. Results

For the input parameters shown in Tables 1 and 2, our test-particle model yields N^+ ion pick up loss rates of about 5×10^{28} s^{-1} (7 XUV), 5×10^{29} s^{-1} (10 XUV) and 2×10^{30} s^{-1} (20 XUV). Table 3 summarizes the integrated atmospheric loss in units of bar over 50 Ma, 100 Ma and 1 Ga.

It shows that if the Earth-analogue test planet would be exposed longer to the various XUV fluxes, the loss accumulates to much higher values. Thus, the stability of the nitrogen reservoirs of Earth-like exoplanets is in greater danger if the nitrogen-rich thermosphere is exposed to higher XUV fluxes, denser stellar winds, or a combination of both.

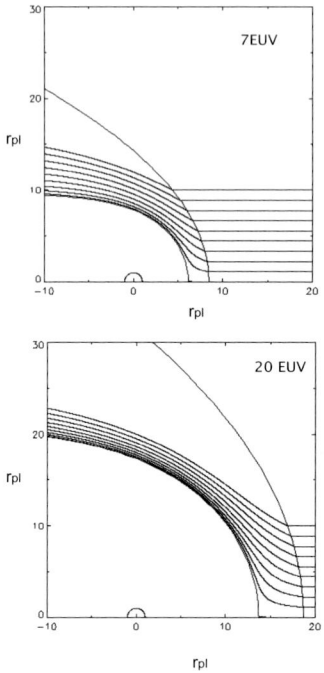

Figure 2. Flow lines, obstacle boundary and bow shock for two different XUV cases. Upper panel (7 XUV case): obstacle corresponds to the magnetopause and is well above the exobase. Lower panel (20 XUV case): obstacle is at the exobase level. The dimensions are given in planetary radii; the small half-circle about the origin represents the planet.

Δt [Ma]	L_7 XUV [bar]	L_{10} XUV [bar]	L_{20} XUV [bar]
50	0.35	3	15
100	0.7	6	30
1000	7	60	300

Table 3. Time integrated nitrogen loss in units of bar for an Earth-analogue planet with a similar atmosphere as present Earth as function of stellar XUV and the parameters given in Tables 1 and 2.

Both stellar factors, high XUV fluxes and dense plasma winds are expected at the orbital locations of M star HZs, therefore our results indicate that the expansion of the upper atmospheres of Earth-analogue nitrogen-rich exoplanets would result in high non-thermal escape rates which most likely remove the initial nitrogen reservoirs from terrestrial planets. One should note that we estimated the N^+ ion pick up loss rates by assuming a fast rotating planet with a strong intrinsic magnetic moment (stronger than that of present Earth).

From these preliminary results we suggest that stable atmospheres of terrestrial exoplanets around XUV active host stars should contain a high mixing ratio of molecules like CO_2, NO, or H_3^+ which can cool the thermospheres due to IR-cooling. Cooler thermospheres will result in less expanded upper atmospheres (see also Lammer et al. 2007) so that the atmosphere can be protected in case the planet remains geodynamically active and generates a strong enough intrinsic magnetic field.

4. Conclusion

High XUV radiation of active M-type stars result in considerable expansion of nitrogen-rich thermospheres of Earth-like exoplanets orbiting their HZ's. Our study shows that even intrinsic magnetic fields comparable or even larger than those of the present Earth can not effectively protect the expanding upper atmosphere against stellar winds or CME plasmas if the XUV flux is ≥ 10 times that of the present solar value. Furthermore, in absence of IR-cooling species like CO_2, NO, or H_3^+ molecules in the thermospheres of such exoplanets, their initial nitrogen reservoirs are in great danger and can be lost via non-thermal atmospheric escape. Non-magnetized or weakly magnetized Earth-like exoplanets which are exposed to high XUV fluxes and stellar plasma flows within the HZ of dwarf stars can be much easier stripped of their whole atmospheres. Our results are in agreement with Lammer et al. (2007) that we do not expect that M-star Earth-like mass or size exoplanets evolve atmospheres and biospheres "analogous" to that of the present Earth even if they orbit within the HZ's.

Acknowledgments. H. Lammer, H. I. M. Lichtenegger, M. L. Khodachenko and Yu. N. Kulikov acknowledge the support of this study by the Helmholtz Association through the research alliance "Planetary Evolution and Life". The authors also thank the International Space Science Institute (ISSI, Bern) because the study was carried out under the framework of the ISSI Team "Evolution of Exoplanet Atmospheres and their Characterization". The authors also thank F. Tian for providing the exobase parameter values from Tian et al. (2008) for the 7 XUV, 10 XUV and 20 XUV cases.

References

Audard, M., Güdel, M., Drake, J.J., Kashyap, V.L. 2000, ApJ, 541, 396
Anicich, V.G., 1993, ApJS, 84, 215
Barth, C.A., Stewart, A.I.F., Bougher, S.W., Hunten, D.M., Bauer, S.J., Nagy, A.F. 1992, in Mars eds. H.H. Kieffer, B.M. Jakosky, C.W. Snyder, & M.S. Matthews, (University of Arizona Press), 1054
Chamberlain, J.W. 1963, Planet. Space. Sci., 11, 996
Erkaev, N. V., Penz, T., Lammer, H., Lichtenegger, H.I.M., Wurz, P., Biernat, H.K., Griessmeier, J.-M., Weiss, W.W. 2005, ApJS, 157, 396
Fox, J.L., Dalgarno, A. 1979, J. Geophys. Res., 84, 7315
Grießmeier, J.-M., Stadelmann, A., Motschmann, U., Belisheva, N.K., Lammer, H., & Biernat H.K. 2005, Astrobiology, 5, 587
Kasting J.F. 1988, Icarus, 74, 47
Khodachenko, M.L., Ribas, I., Lammer, H., Grießmeier, J.-M., Leitner, M., Selsis, F., Eiroa, C., Hanslmeier, A., Biernat, H.K., Farrugia, C.J., & Rucker, H.O. 2007, Astrobiology, 7, 167
Kulikov, Yu.N., Lammer, H., Lichtenegger, H.I.M., Terada N., Ribas, I., Kolb, C., Langmayr, D., Lundin, R., Guinan, E.F., Barabash, S., & Biernat, H.K. 2006, Planet. Space Sci., 54, 1425
Lammer, H., Lichtenegger, H.I.M., Kolb, C., Ribas, I., Guinan, E.F., Bauer, S.J. 2003, Icarus, 165, 9

Lammer, H., Lichtenegger, H.I.M., Kulikov, Yu.N., Grießmeier, J.-M., Terada, N., Erkaev, N.V., Biernat, H.K., Khodachenko, M.L., Ribas, I., Penz, T., & Selsis, F. 2007, Astrobiology, 7, 185

Nagy, A.F., Cravens, T.E., Gombosi, T.I. 1983, in Venus, ed. D.M. Hunten, L. Colin, T.M. Donahue, and V.I. Moroz, (University of Arizona Press), 841

Ribas, I., Guinan, E.F., M. Güdel, & Audard, M. 2005, ApJ., 622, 680

Roble, R.G. 1995, Geophys. Monogr. Ser., 87, ed. R.M. Johnson & T.L. Killeen, (AGU, Washington D.C.), 121

Scalo, J., Kaltenegger, L., Segura, A., Fridlund, M., Ribas, I., Kulikov, Yu.N., Grenfell, J.L., Rauer, H., Odert, P., Leitzinger, M., Selsis, F., Khodachenko, M.L., Eiora, C., Kasting, J., & Lammer, H. 2007, Astrobiology, 7, 85

Smithtro, C.G., Sojka, J.J. 2005a, J. Geophys. Res., 110, DOI:10.1029/2004JA010781

Smithtro, C.G., Sojka, J.J. 2005b, 110, DOI:10.1029/2004JA010782

Tian, F., Kasting, J.F., Liu, H., Roble, R.G. 2008, J. Geophys. Res., 113, DOI:10.1029/2007JE002946

Yamauchi, M., Wahlund J.-E. 2007, Astrobiology, 7, 783

Molecules in the Atmospheres of Extrasolar Planets
ASP Conference Series, Vol. 450
J.-P. Beaulieu, S. Dieters, and G. Tinetti, eds.
© 2011 Astronomical Society of the Pacific

EXOFIT: Bayesian Estimation of Orbital Parameters of Extrasolar Planets

Sreekumar T. Balan

Astrophysics Group, Cavendish Laboratory, JJ Thomson Avenue, Cambridge CB3 0HE, UK

Ofer Lahav

Department of Physics and Astronomy, University College, Gower Street, London WC1E 6BT, UK

Abstract. We introduce EXOFIT, a Bayesian tool for estimating orbital parameters of extra-solar planets from radial velocity measurements. EXOFIT can search for either one or two planets at present. EXOFIT employs Markov Chain Monte Carlo method implemented in an object oriented manner. As an example we re-analyze the orbital solution of HD155358b and the results are compared with that of the published orbital parameters. In order to check the agreement of the EXOFIT orbital parameters with the published ones we examined radial velocity data of 30 stars taken randomly from *www.exoplanet.eu*. We show that while orbital periods agree in both methods, EXOFIT prefers lower eccentricity solutions for planets with higher ($e \geq 0.5$) orbital eccentricities.

1. Introduction

More than a decade of extensive search for extra-solar planets has resulted in nearly 300 planets. Majority of the contribution to the extra-solar planet count comes from radial velocity method. Radial velocity data is traditionally analyzed first by a periodogram (Lomb, 1976, Scargle 1982) for the orbital period and then for the other orbital parameters using conventional optimization methods. Bayesian methods for the estimation of orbital parameters of extra-solar planets were introduced by Gregory (2005) and ford (2005). Their work shows that these methods provide us a framework to tackle the problems associated with the traditional methods in a transparent and robust manner. EXOFIT[1] of Balan & Lahav (2008) is the first publicly available package for Bayesian estimation orbital parameters from radial velocity measurements. In this article we discuss the application of EXOFIT to the radial velocity data of HD155358 and compare our results with the published orbital solution (Cochran et al., 2007). We also show, by analyzing a randomly selected sample of radial velocity data from *www.exoplanet.eu* that EXOFIT disagrees with the published orbital eccentricities in many instances while the orbital periods from EXOFIT matches closely with the published ones.

---

[1] www.star.ucl.ac.uk/~lahav/exofit.html

The rest of the article is organized as follows. In Section 2 we give a brief introduction to radial velocity modeling. Section 3 describes Bayesian approach to parameter estimation. EXOFIT is introduced in Section 4 and its application to the radial velocity data is discussed in section 5. Section 6 explains the analysis of radial velocity data of a sample of stars taken form *www.exoplanet.eu* using EXOFIT. We conclude this article in Section 7 and provide an outline on planned work.

2. Modeling of Radial Velocity Data

Radial velocity data consists of a set of measured radial velocity entries, corresponding time of observation and the uncertainty in each measurement. Observed radial velocity data is modeled by the equation

$$d_i = v_i + \epsilon_i + \delta, \qquad (1)$$

where d_i is the measured radial velocity data for the ith instant of time t_i, v_i is true radial velocity of the star, ϵ_i is the measurement error assigned by the observer and δ represents any unknown noise present (e.g., signal from another planet) in the data (Gregory, 2005). For a statistician, δ is a nuisance parameter. The true radial velocity can be simulated using a mathematical model. Disregarding any interactions between planets, the radial velocity of a star for a typical n-planet model can be approximately written as a linear combination of n single planet radial velocities. Thus,

$$v = V - \sum_{i=1}^{n} K_i (\sin(f_i + \varpi_i) + e_i \sin \varpi_i), \qquad (2)$$

where V, K, f, e, w represent the systematic velocity, amplitude, true anomaly, eccentricity and the longitude of periastron respectively. For the full formalism see the User's Guide to EXOFIT.

3. Bayesian Parameter Estimation

Bayesian paradigm has its origins in an article published posthumously by Rev. Thomas Bayes in 1763. Since then the theorem has played a central part in probabilistic inference. For latest examples in cosmology see Feroz & Hobson (2008) and Lewis & Bridle (2002). Bayesian methods for the estimation of orbital parameters of extra-solar planets were introduced by Gregory (2005) and Ford (2005) and their research show that this approach has an edge over the traditional methods when dealing with for e.g., highly eccentric orbits. These methods also provide a straight forward way of dealing with nuisance parameters and a robust way of estimating uncertainties associated with the estimates of orbital parameters.

Bayes' theorem for a set of parameters Θ in model H and data $\mathbf{D}$ can be written as,

$$\Pr(\Theta|\mathbf{D}, H) = \frac{\Pr(\mathbf{D}|\Theta, H) \Pr(\Theta|H)}{\Pr(\mathbf{D}|H)}. \qquad (3)$$

In the above equation $\Pr(\Theta|\mathbf{D}, H)$ is the posterior probability distribution of parameters, $\Pr(\mathbf{D}|\Theta, H)$ is the likelihood of the data, $\Pr(\Theta|H)$ represents the prior probability distribution of the parameters and $\Pr(\mathbf{D}|H)$ is called the Bayesian evidence. For parameter estimation problems we could simply write the above equation as

$$\Pr(\Theta|\mathbf{D}, H) \propto \Pr(\mathbf{D}|\Theta, H)\Pr(\Theta|H) \tag{4}$$

Computing the right hand side of Equation 4 is the central point of any Bayesian parameter estimation. Analytical solutions can be derived for some special cases. However, in general we use numerical methods to compute the posterior distribution. Although many approximation methods exist, this area is dominated by Markov Chain Monte Carlo (MCMC) and other sampling methods. These rely on the a random walk through the parameter space and make use of the fact that posterior density is proportional to the number of points visited in the volume considered. We calculate the marginal posterior distribution of each parameter by simple plotting a histogram of the final set of samples.

Bayesian modeling of the problem consists of defining each of the components mentioned in Equation 4. This will be discussed from the context of estimation of orbital parameters when we apply EXOFIT to the published radial velocity data of HD155358 in Section 5.

4. EXOFIT

EXOFIT is an easy to use documented software for estimating the orbital parameters from radial velocity measurements and it is freely available. It is based on Bayesian MCMC method and is implemented in an object oriented framework in C++. It can be easily extended to analyze the radial velocity data of more than two planets as well as data from transit photometry. Output of EXOFIT is a set of posterior samples. These can be analyzed using any standard statistical software. This explained in User's Guide which can be download from EXOFIT website[2]. We also provide sample script for R statistical environment[3] to analyze posterior samples. Improved parameterization for the problem and novel sampling techniques for EXOFIT are under development.

5. Application to Radial Velocity Data of HD155358

In this section we develop a Bayesian model for the parameter estimation and apply EXOFIT to extract the orbital parameters of the companions of HD155358. We start by defining the likelihood of the data. As mentioned in Section 2 we assume that the data consists of true radial velocity and some noise. The component of noise arising from the known measurement errors ϵ_i is assumed to be normally distributed with standard deviation σ_i for the ith entry in the data. The probability distribution for the unaccounted noise component δ is chosen to a Gaussian distribution with finite variance s^2. Therefore, the distribution of the combination $\epsilon_i + \delta$ can be considered as a Gaussian with a variance of $\sigma_i^2 + s^2$.

Assuming each measurement error ϵ_i to be independent and since they follow a Gaussian distribution, the likelihood of data can be written as a product N Gaussians

---

[2] www.star.ucl.ac.uk/~lahav/exofit.html

[3] www.r-project.org

(Gregory, 2005) where N is the number of entries in the data. Therefore,

$$\Pr(\mathbf{D}|\Theta, H) = A \, \exp\left[-\sum_{i=1}^{N} \frac{(d_i - v_i)^2}{2(\sigma_i^2 + s^2)}\right], \tag{5}$$

where

$$A = (2\pi)^{-N/2}\left[\prod_{i=1}^{N}(\sigma_i^2 + s^2)^{-1/2}\right]. \tag{6}$$

Thus, our parameter space is $\{V, T_1, K_1, e_1, w_1, \chi_1, T_2, K_2, e_2, w_2, \chi_2, s\}$, first 11 from the mathematical model and last one representing the nuisance parameter δ. The parameter is χ is defined for computational purposes and marks the periastron passage time as function period T. For more details please consult EXOFIT User's Guide.

Table 1, taken from Balan & Lahav (2008) gives the prior probability distributions of each parameter in the model. These priors are chosen in such a way that they allow likelihood term in the Equation 4 to dominate the posterior distribution and thus ensuring inference to be drawn from the observed data.

Table 1. The assumed prior distribution of orbital parameters and their boundaries for a 2-planet model.

Para.	Prior	Mathematical Form	Min	Max
$V(ms^{-1})$	Uniform	$\frac{1}{V_{max}-V_{min}}$	-2000	2000
$T_1(days)$	Jeffreys	$\frac{1}{T_1 \ln\left(\frac{T_{1max}}{T_{1min}}\right)}$	0.2	15000
$K_1(ms^{-1})$	Mod. Jeffreys	$\frac{(K_1+K_{10})^{-1}}{\ln\left(\frac{K_{10}+K_{1max}}{K_{10}}\right)}$	0.0	2000
e_1	Uniform	1	0	1
ϖ_1	Uniform	$\frac{1}{2\pi}$	0	2π
χ_1	Uniform	1	0	1
$T_2(days)$	Jeffreys	$\frac{1}{T_2 \ln\left(\frac{T_{2max}}{T_{2min}}\right)}$	0.2	15000
$K_2(ms^{-1})$	Mod. Jeffreys	$\frac{(K_2+K_{20})^{-1}}{\ln\left(\frac{K_{20}+K_{2max}}{K_{20}}\right)}$	0.0	2000
e_2	Uniform	1	0	1
ϖ_2	Uniform	$\frac{1}{2\pi}$	0	2π
χ_2	Uniform	1	0	1
$s(ms^{-1})$	Mod. Jeffreys	$\frac{(s+s_0)^{-1}}{\ln\left(\frac{s_0+s_{max}}{s_0}\right)}$	0	2000

Application of EXOFIT revealed the posterior distribution of planets as given in Figure 1(a) and the corresponding radial velocity curve is shown in Figure 1(b). We compare our results to the published results by Cochran et al. (2007) in Table 2. Although our results looks similar, we notice a noise factor(s) of $5.43\,ms^{-1}$ which indicates the presence of an additional signal in the data. We did look for a third planet in the system, but our results were inconclusive. This issue can be settled if we have more observations down the line.

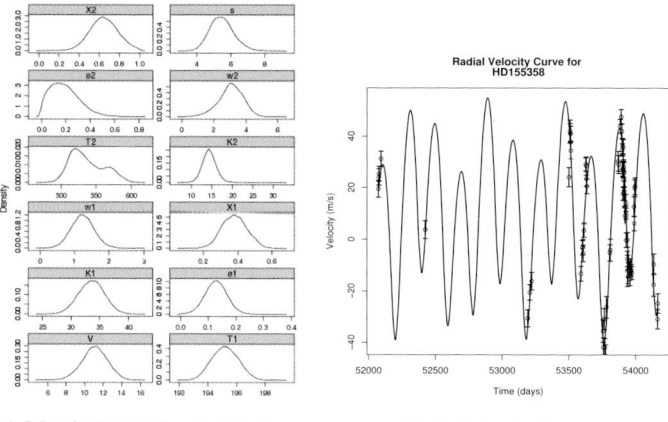

(a) Marginal posterior destributions (b) Radial velocity curve

Figure 1. Marginal postrior distribuotions of orbital orbital parameters of the companions to HD155358 are shown in the left panel. Right panel shows the corresponding radial velocity curve for HD155358.

6. EXOFIT vs exoplanet.eu

In this section we compare the published orbital periods and eccentricities of 30 extrasolar planets taken randomly from *www.exoplanet.eu* to that obtained by analyzing corresponding radial velocity data with EXOFIT. The results are summarized in Figure 2. It can be observed that the orbital periods extracted with EXOFIT matches closely with published ones. However, the eccentricities show apparent variation from the published values. In fact EXOFIT tends to obtain lower eccentricity solutions. This is clearly noticeable for planets with eccentricities greater that 0.5 (i.e. $\simeq$ 10% of the planet population). Median was used as the point estimator for marginal posterior distribution orbital parameters of exo-planets extracted with EXOFIT while comparing with the published results. This suggests that the orbital eccentricity is poorly constrained in many occasions because of the sparse sampling of the data points.

7. Summary and Future Work

We have introduced EXOFIT by reanalyzing the orbital solution of HD155358 by Cochran et al. (2007). EXOFIT provides a full Bayesian analysis of the problem and spits out the marginal posterior distribution of orbital parameters. We have also compared the orbital parameters obtained by EXOFIT to the published orbital solutions of 30 extra-solar planets taken randomly from *www.exoplanet.eu*, the objective being the reanalysis of the data using a single method and the estimation of corresponding orbital parameters and their uncertainties. The results show that while the orbital periods from EXOFIT agrees closely with the published ones, the eccentricities show significant variation from the published results. This fact indicates that orbital eccentricity is not accurately constrained and this degeneracy should be considered more carefully.

Table 2. A table showing the summary of the posterior distribution of orbital parameters extracted with EXOFIT 2-planet model and the published orbital solution by Cochran et al. (2007). Columns 2, 3 and 4 show the posterior mean (and standard deviation), median (and 25% and 75% quantiles) and the maximum a posteriori, i.e. posterior mode (and 68.3% highest density regions) respectively. Mass of the star was assumed to be $0.87 M_{\oplus}$.

Parameters	EXOFIT(Mean)	EXOFIT(Median)	EXOFIT(Mode)	Cochran et al.
$V(ms^{-1})$	11.09 ± 1.37	$11.11^{+0.91}_{-0.91}$	$11.23^{+1.17}_{-1.32}$	
$T_1(days)$	195.22 ± 0.91	$195.20^{+0.64}_{-0.63}$	$195.17^{+0.87}_{-0.84}$	195 ± 1.1
$K_1(ms^{-1})$	33.56 ± 2.22	$33.59^{+1.48}_{-1.55}$	$33.58^{+2.07}_{-2.02}$	34.6 ± 3
e_1	0.13 ± 0.04	$0.13^{+0.03}_{-0.03}$	$0.13^{+0.03}_{-0.04}$	0.112 ± 0.037
$\varpi_1(degrees)$	160.78 ± 19.73	$160.46^{+12.82}_{-12.77}$	$158.20^{+19.30}_{-15.33}$	162 ± 20
χ_1	0.38 ± 0.07	$0.38^{+0.05}_{-0.05}$	$0.38^{+0.06}_{-0.07}$	
$T_2(days)$	537.05 ± 24.00	$532.15^{+22.78}_{-13.84}$	$519.77^{+29.19}_{-13.46}$	530.3 ± 27.2
$K_2(ms^{-1})$	14.40 ± 1.68	$14.34^{+1.07}_{-1.03}$	$14.21^{+1.52}_{-1.35}$	14.1 ± 1.6
e_2	0.19 ± 0.12	$0.18^{+0.09}_{-0.08}$	$0.14^{+0.11}_{-0.10}$	0.176 ± 0.174
$\varpi_2(degrees)$	265.62 ± 45.42	$267.53^{+28.16}_{-28.62}$	$265.28^{+45.48}_{-32.34}$	279 ± 38
χ_2	0.64 ± 0.14	$0.64^{+0.10}_{-0.09}$	$0.63^{+0.14}_{-0.11}$	
$s(ms^{-1})$	5.47 ± 0.69	$5.44^{+0.47}_{-0.45}$	$5.48^{+0.52}_{-0.70}$	
$Tp_1(BJD)$	2453946.95 ± 13.93	$2453946.95^{+10.00}_{-10.00}$	$2453950.63^{+14.26}_{-13.90}$	2453950 ± 10.4
$M_1 \sin i$	0.87 ± 0.06	$0.87^{+0.04}_{-0.04}$	$0.87^{+0.06}_{-0.06}$	0.89 ± 0.12
$a_1(AU)$	0.63 ± 0.00	$0.63^{+0.00}_{-0.00}$	$0.63^{+0.00}_{-0.00}$	0.628 ± 0.02
$Tp_2(BJD)$	2454408.22 ± 70.22	$2454408.22^{+50.00}_{-50.00}$	$2454420.27^{+60.57}_{-76.26}$	2454420.3 ± 79.3
$M_2 \sin i$	0.51 ± 0.06	$0.51^{+0.04}_{-0.04}$	$0.51^{+0.06}_{-0.06}$	0.504 ± 0.075
$a_2(AU)$	1.23 ± 0.04	$1.23^{+0.03}_{-0.02}$	$1.21^{+0.07}_{-0.02}$	1.224 ± 0.081

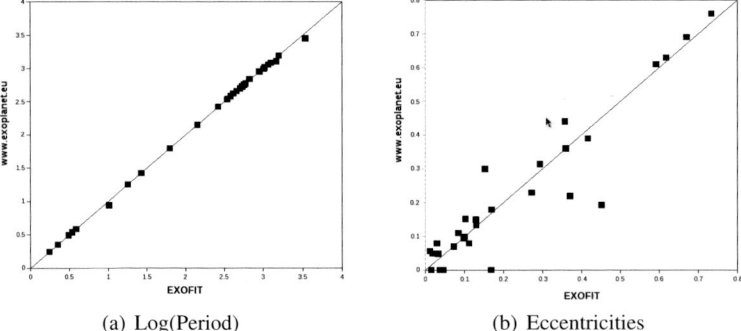

(a) Log(Period)　　　　(b) Eccentricities

Figure 2. Figure on the left plots log(orbital period) of extra-solar planets, with results obtained by EXOFIT on the horizontal axis and published ones on the vertical axis. Figure on the right shows a comparison of orbital eccentricities with results obtained from EXOFIT on the horizontal axis and published ones on the vertical axis.

We plan to improve to the efficiency of EXOFIT by considering new parameterization for the problem as well as faster sampling techniques. Bayesian model selection will also be considered for the future versions of EXOFIT. We intend to extend our analysis to more planets and planetary data from transit photometry to provide a comprehensive Bayesian analysis of the statistical properties of orbital parameters of extra-solar planets.

Acknowledgments. SB and OL would like to thank the organizers of the *Molecules in extrasolar planet atmosphere conference* for giving the opportunity to attend the conference. SB would like the thank the organizing committee for the financial support provided. OL acknowledges the support of a Royal Society Wolfson Research Merit Award.

References

Balan, S. T., & Lahav O., 2008, MNRAS, 394, 1936
Cochran, W. D., Endl, M., Wittenmyer, R. A., Bean, J. L., 2007, ApJ, 665, 1407
Feroz F., Hobson, M. P., 2008, MNRAS, 384, 449
Ford E. B, 2005, AJ, 129, 1706
Gregory, P. C. , 2005, ApJ, 631, 1198
Jones, H. R. A., Butler, R. P., Tinney, C. G., Marcy, G. W., Carter, B. D, Penny, A. J., McCarthy, C., & Bailey J., 2006, MNRAS, 369, 249
Lewis, A., Bridle, S., 2002, Phys.Rev.D, 66, 103511
Lomb, N. R. , 1976, Ap&SS, 39, 447
Scargle, J. D. , 1982, ApJ, 263, 835

Enric Palle reporting observations of the dark side of the moon.

What could be observed in the case of Super-Ios and Hyper-Ios?

D. Briot, E Lellouch and J. Schneider

Observatoire de Paris-Meudon

Abstract. Since the detection of a very strong volcanic activity on Io by Voyager, other instances of volcanic activity have been discovered in the solar system. Super-Earths are very promising extra-solar planets. If a volcanic activity is generated in a similar manner as for Io occurs in Super-Earth's, these objects would be highly active corresponding to Super-Io's and even more Hyper-Ios. They would present very interesting observational features. We investigate the conditions of a possible atmosphere and its properties, particularly the molecules corresponding to a volcanic activity.

1. Introduction

The discovery of a volcanic activity of Io by Voyager spacecraft in 1979 (Morabito et al., 1979) was predicted by Peales et al. (1979). Since then, other discoveries concerning several bodies of the solar system definitely showed that a volcanic activity is probably a current event for planets and satellites. In the solar system, a volcanic activity is observed in the larger "solid" body, as well as in some satellites, demonstrating that this process can happen in bodies of various sizes and masses. However, the volcanic activity of Io is very special because of its intensity. Detection and study of a volcanic activity on a terrestrial planet or a Super-Earth would be really interesting, because in this case the luminosity and the calorific energy produced by the planet change observational conditions. If we imagine a Super-Earth or a terrestrial planet with a volcanic activity as intense as Io's, the conditions of observations of this Super Earth would be greatly modified and improved,
- in case of transit
- in case of a secondary transit
- by imaging.

We shall call such a planet a SUPER-IO, in case of a terrestrial mass and even higher level a HYPER-IO in case of a mass corresponding to a super-Earth.

An other interesting point would be that heating produced by the planet would allow to enlarge and precise the definition of the habitability zone, which is generally defined only as a function of both the star temperature and the distance of the planet

2. Possibility of volcanic activity in terrestrial planets and Super-Earths

The volcanism of Io is attributed to the tidal forces due to Jupiter. These forces are important because of both the closeness of Jupiter and Io, and the eccentricity of the Io orbit. Io's orbit is made eccentric by resonance with orbits of Europa and Ganymede.

Actually, extrasolar planets are frequently very close to their star, their trajectories are often eccentric, and in case of a system of several planets, cases of resonance of orbits are frequently observed. The system of Jupiter and its larger satellites maybe exists on a much larger scale in some stars and their planetary system. So the probability that there are is volcanic activity similar to Io's one in terrestrial extra-solar planets or in Super-Earths looks very high. It appears that a high proportion of Super-Earths can possibly display volcanic activity, and even a very strong volcanic activity (see e.g. Jackson et al., 2008)

3. Some physical properties of Super-Ios and Hyper-Ios

Because of its vicinity, Io is very well studied and its physical properties are well known. This knowledge can be used to imply properties of SuperIos and HyperIos. However some physical conditions are quite different in Io and in extrasolar planets with very intense activity. Let us note that like Io around Jupiter, Super-Ios likely would be co-rotating around their star.

3.1. Temperature

Io is rotating closely around a planet and is remote from its star. A super-Io is to be close its star and so much hotter, at least its hemisphere facing the star, in which an extreme heat can be presumed. Temperature in the other hemisphere would depend on physical conditions of any atmosphere.

3.2. Mass

Io's mass is around 0.015 Earth mass while the smallest masses of detected planets are around the terrestrial mass. Many years will probably be necessary before an like-Io mass object, planet or satellite, could be detected. The atmosphere of a Super-Io, or a Hyper-Io would likely be much thicker than Io's atmosphere. This atmosphere is fed by material ejected by volcanoes and would contain sulphur dioxide.

4. Observations of Super-Ios and Hyper-Ios

In case of a strong volcanic activity in a Earth or a Super-Earth, observations would be very special. Some observational features can be deduced from Io's observations, but, as said above, there are also very strong differences :

- Because of the vicinity of the star, instead of Jupiter, the temperature on a Super-Io would be much higher than on Io.
- The mass of the objects of the system studied here i.e. star and Super-Io, are much bigger than in the system of Jupiter and Io.

4.1. Transit of a Super-Io or a Hyper-Io

As said above, the much larger size of a Super-Earth with a volcanic activity, as compared with the real Io, implies that a Super-Io would have a very much thicker atmosphere than Io. The spectrum of the sulphur dioxide would be observed in case of transit and would be a signature of volcanic activity.

4.2. Secondary Transit of a Super-Io or a Hyper-Io

The volcanic activity of Io is specially important in the near infrared, in the $3 - 5\mu$ m range. The self radiation of a Super-Io makes it as an excellent candidate for observations of secondary transits at these wavelengths, especially in the case of a cool star. The volcanic activity of Io appears in hot spots. The volcanoes of Io can be detected in the case of an occultation by Europa (see Descamps et al., 1992). The hot spots of a Super-Io could be detected in the first part and the last part of the secondary transit, if some small irregularities are observed during the decrease and the increase of the luminosity. If this kind of small irregularities could not be detected, it could be a indication of a very opaque atmosphere, preventing the observation of individual volcanoes.

4.3. Visibility of a Super-Io or a Hyper-Io by Imaging

If a Super-Io emits its own radiation, the ratio between the luminosity of the star and the luminosity of the planet, specially in the case of a cool star, becomes smaller than in the case of a "inertial" i.e. a "non-volcanic" planet and a solar type star. So these hypothetical objects would become very interesting objects to be observed by imaging in the infrared.

5. Conclusion

Tidal effects could be very important in extrasolar planets and occurrence of extrasolar planets undergoing volcanic activity similar in nature to Io but on a larger scale is truly possible. These super-Ios and hyper-Ios would be both much hotter and more massive than Io. They would be very interesting objets to detect, to observe, and to study. The spectrum of sulphur dioxide would be detected. The physical conditions of these objects could be extreme and radically different from any known object. So, the observational features would be very special and have to be taken into account in development of future instrumental projects.

Acknowledgments. It is a pleasure to thank Monique Spite and Frédéric Royer for their very helpful advice.

References

Descamps, P., Arlot, J.E., Thuillot, W., Colas, F., & Vu, D.T. 1992, Icarus, 100, 235
Jackson, B., Barnes, R., & Greenberg, R. 2008, MNRAS, 391, 237
Morabito, L.A., Synott, S.P., Kupferman, P.N., & Collins, S.A. 1979, Science, 204, 972
Peale, S.J., Cassen, P., & Reynolds, R.T. 1979, Science, 203, 892

Note added in proof

Volcanism on extrasolar planets recently became a subject of real interest, (see e. g. Kaltenegger, Henning, & Sasselov, 2010. Some newly discovered planets are now studied in the frame of this hypothesis. The very interesting extrasolar planet CoRoT-7b is a good example (Barnes et al., 2010).

References

Barnes, R., Raymond, S.N., Greenberg, R., Jackson, B., & Kaib, N.A. 2010, ApJL, 709, L95
Kaltenegger, L., Henning, W.G., & Sasselov, D.D. 2010, Astron. J. , 140, 1370

Vincent Coude du Foresto: session chairman and representing the Blue Dot Team.

Part VII

The Future: Short and Long Term Missions and Instruments to Characterise Exoplanet Atmospheres. (Chair : Vincent Coude du Foresto)

Direct Imaging of Extrasolar Planets: Overview of Ground and Space Programs

A. Boccaletti

Laboratoire d'Etude Spatiale et d'Instrumentation en Astronomie, Observatoire de Paris, F-92195 Meudon, France

Abstract. With the ever-growing number of exoplanets detected, the issue of characterization is becoming more and more relevant. Direct imaging is certainly the most efficient but the most challenging tool to probe the atmosphere of exoplanets and hence in turns determine the physical properties and refine models of exoplanets. A number of instruments optimized for exoplanets imaging are now operating or planned for the short and long term both on the ground and in space. This paper reviews these instruments and their characteristics/capabilities. Conclusions are drawn on the spectral characterization point of view.

1. Context

The study of extrasolar planets has became in a decade an exciting field in modern astronomy. With more than 300 planets detected so far, indirect methods have been the most prolific at finding sub-stellar objects in the solar neighborhood. However, a few angularly resolved images of Extrasolar Giant Planets (EGPs) have been finally obtained with 8-10m class telescopes on the ground as well as in space with the HST (Chauvin et al. 2005; Neuhauser et al. 2005; Lafrenière et al. 2008, for instance,). The detection rate is progressing quickly. Some discoveries were announced a week or so before the conference (Kalas et al. 2008; Marois et al. 2008; Lagrange et al. 2008). These outstanding observations were made possible by the favorable configurations of these planetary systems: low star/planet mass ratio, young age (a few tens to hundreds Myr), large physical distances (typically hundreds of AU, except for β Pic) which basically make these planets bright enough or locally above the stellar halo. A list of objects detected by direct imaging is summarized in Tab. 1.

This paper presents an overview of direct imaging projects on single apertures. As guideline for non-specialists, the discussion remains general and more details can be found in reference papers. The first part briefly reminds the problematic of high contrast imaging and the instrumental solutions that are being developed. In section 3 and 4 the projects are separated in two categories: those that are planned and those that are proposed for the future and mostly focused on telluric planets. The intent is to clearly points out what is going to be achieved in terms of parameter space in order to better define what will be needed for the next projects. Finally, all these projects are sorted along a timeline and some conclusions are derived.

Table 1. List of planetary mass objects detected with direct imaging, as of Feb. 2009. Error bars are not indicated but are often larger than 1 M_J therefore placing some of these objects in the Brown Dwarf regime.

object	estimated mass [M_J]	estimated separation [AU]	reference
2M1207 b	5	46	Chauvin et al. (2005)
GQ Lup b	17	100	Neuhauser et al. (2005)
AB Pic b	14	248	Chauvin et al. (2005)
CHRX73 b	12	210	Luhman et al. (2006)
HN Peg b	16	795	Luhman et al. (2007)
DH Tau b	12	330	Itoh et al. (2005)
RSX 1609 b	8	330	Lafrenière et al. (2008)
Fomalhaut b	3	120	Kalas et al. (2008)
HR8799 b	5	46	Marois et al. (2008)
HR8799 c	12	330	Marois et al. (2008)
HR8799 d	8	330	Marois et al. (2008)
β Pic b	8	8	Lagrange et al. (2008)

2. Problems and Solutions

The problematic is well known: planets are much fainter than parent stars and angularly close. The Sun-Jupiter example is often cited as a reference, with a contrast of about 10^9 and a angular separation of 0.5" at 10pc. However, the realm of planets exhibits a much wider variety. The star to planet brightness ratio strongly depends on the characteristics of the system (age, temperature, physical distance, radius) and the spectral range. Actual contrasts of planets range between 10^4 and 10^{10}.

The huge contrast issue is emphasized by the diffraction which makes the stellar light extending all across the focal plane. As a consequence, the starlight overshines the planet light and produces a large photon noise. A system to remove or attenuate the starlight is therefore mandatory to improve the signal to noise ratio at the planet position in the field. On single aperture the solution is to use a coronagraphic system as Lyot did in the 1930's to observe the solar corona (Lyot 1939). But, the big difference with solar observations is that stars are point like source and therefore images are dominated by diffraction. Since 1996 (Gay & Rabbia 1996) alternative concepts have been studied. Large rejections have been searched for and achromaticity have been considered as a major issue. To date, many coronagraph concepts do exist. An almost exhautive list is given in Guyon et al. (2006) and a classification in Quirrenbach (2005). Also, many of them were prototyped and tested successfully at large contrast (Baudoz et al. 2007, for instance). A few were also implemented on current telescopes (Boccaletti et al. 2004). The interesting point in this race towards high contrast is that many concepts of coronagraph provide a perfect attenuation of the starlight in some particular conditions (shape of pupil, bandwidth, ...), at least on the paper. Prototyping activities have demonstrated that the perfect starlight rejection can be approached but never reached. Nevertheless, some concepts have been elaborated to a sufficiently high level that is compatible with planet detection.

The second issue, is related to wavefront aberrations. Even a perfect coronagraph only attenuates the coherent part of the wavefront. Aberrations are leaking through coronagraphs and produce a residual intensity in the focal plane in the form of a speckled halo. Two families of techniques are considered to tackle the speckle issue. The first one is more straightforward and is using wavefront correction and therefore necessitates a measurement of this wavefront and one or more corrective elements (usually a deformable mirror). Several testbeds are developed to address this problem of wavefront correction. Technical implementations mostly differ by the algorithm used to measure the wavefront like speckle minimization for instance (Bordé & Traub 2006; Givcon et al. 2008), while all testbeds consider measurements in the coronagraphic image plane and deformable mirror to apply the correction. Very high contrasts have been already demonstrated in the lab with wavefront correction (Trauger & Traub 2007). Other designs are being studied (Galicher et al. 2008; Codona & Angel 2004).

An alternative to wavefront correction is speckle calibration, the idea being to disentangle the stellar speckles from the planet peak owing to particular characteristics. For instance, planets have spectral or polarimetric signatures not present in the starlight. On the ground, simultaneous spectral (Racine et al. 1999) or polarimetric measurements (Kuhn et al. 2001) are required to get rid of the atmospheric residual speckles as of static speckles. The so-called differential imaging techniques might be affected by defects inherent to the method (like chromatism for spectral differential imaging) or to the instrument (differential aberrations in the optical path). Some algorithms and optical concepts are studied to reduce the impact of these defects like spectral deconvolution (Sparks & Ford 2002). In this respect, Integral Field Spectroscopy is considered as a promising method. Angular differential imaging (Marois et al. 2006) is another sort of speckle calibration technique. With an alt-az ground-based telescope the planet rotates with the field around the star while speckles are either static (if originating to the instrument) or rapidly evolving (if originating from the atmosphere). Aberrations that slowly evolve as the field rotates can still mimic angularly separated companions. Also, the performance is improving with the angular separation while a minimal angle can be defined (a few tenths of arcsec). This technique has recently been successful at finding 3 planets around the same star (Marois et al. 2008). Finally, planets and stars provide incoherent wavefronts and this characteristic can also be exploited to reveal planets amongst speckles (Guyon 2004; Baudoz et al. 2006; Codona & Angel 2004).

The bottom line, is that coronagraph as speckle rejection concepts have been demonstrated. The ability to make these systems efficient enough is now an engineering issue.

3. Planned Projects

3.1. "Planet Finders" on 8-m Class Telescopes (2011)

Ground based 8-m class telescopes are now equipped with Adaptive Optics (AO) systems and some are already including high contrast imaging facilities like coronagraphy (Boccaletti et al. 2004) or differential imaging (Lenzen et al. 2004). A few discoveries of giant planets were already possible as mentioned above in some favorable conditions. To routinely achieve higher contrasts, "planet finder" instruments were planned since 2001-2002 at VLT, Gemini and Subaru. SPHERE (Spectro Polarimetric High contrast Exoplanet REsearch, Berzit et al. 2006), GPI (Gemini Planet Imager, Macintosh et al. 2006) and HiCiAO (High-contrast Coronagraphic Imager with Adaptive Optics, Tamura et al. 2006) are sharing the same conceptual design which is to combine extreme AO, broad band coronagraphy and spectral differential imaging. These instruments slightly differ on the choice of the coronagraphic or differential systems but will certainly deliver very similar performance. SPHERE also extends to the visible range with a differential polarimetric imager to take advantage of the increase of the reflected light for very close planets (Schmid et al. 2006).

Typical targets by order of importance are :

- Young and very young stars: planets in young systems are warm and hence self luminous. Evolutionnary models (Burrows et al. 1997; Chabrier et al. 2000) are predicting higher luminosity than for mature planets by several orders of magnitude depending on age and mass. Adequate targets are a few tens of Myr old and within 100 pc. Spectral differential imaging will be efficient for planets with $300 < T(K) < 1300$.

- Stars with known planets: the improvement of temporal coverage makes possible the detection of long period giants with radial velocity (about 100 planets with P>5 years according to http://exoplanet.eu). Many stars also exhibits long-term residuals indicating massive planets at large separations appropriate for direct imaging.

- Nearby stars (<5 pc): the spectrum of irradiated giant planets at <1 AU has a significant reflected light component that can be detected in the visible owing to the gain in angular resolution.

Large surveys will be necessary with such instruments to discover new planets. Near IR spectra of these planets at low or medium resolution ($50 < R < 800$) will be feasible for the first time. Further details on performance estimate can be found in Marois et al. (2008b); Vigan et al. (2008); Thalman et al. (2008); Boccaletti et al. (2008).

3.2. JWST (2014)

The James Webb Space Telescope is an observatory that allows diffraction limited imaging at wavelengths longer than $2\mu m$. It comes with a suite of 4 instruments out of 3 being equipped with coronagraphs or high contrast facilities designed for the characterization of extrasolar giant planets.

- NIRCAM, the near IR camera (0.6-$5\mu m$) has 5 coronagraphs (Krist 2007) based on the band-limited concept (Kuchner & Traub 2002) and combined with 5-10% bandwidth filters (between 2 and $4.8\mu m$). The best performance are expected at $4.6\mu m$ owing to a peak in the thermal emission of EGPs. At this band the coronagraph is able to look at separation larger than 0.6" and will achieve a contrast level of about $10^5 - 10^6$ (at about 1-2"). This limit of detection (Krist 2007) corresponds to a mass of $2\,M_J$ for 1 Gyr and less than $1\,M_J$ for young objects (<300 Myr).

- MIRI, the mid IR camera (5-$28\mu m$) has 4 coronagraphs (Boccaletti et al. 2005) based on the Four Quadrant Phase Mask concept (Rouan et al. 2000) and combined with 5% bandwidth filters (at 10.65, 11.40 and $15.50\mu m$). Contrasts of

$10^4 - 10^5$ are achievable between 0.5 and 1". This should allow detection of 5 M_J mature planets.

- TFI is a tunable filter imager (1.6-4.9μm) implemented in the fine guidance sensor capable of coronagraphic imaging with a resolution of R=100 (Doyon et al. 2008). Taking advantage of the spectral deconvolution technique it will be able to achieve a contrast of 10^5 at 1" with a 0.6" lyot mask. Therefore, TFI will have similar performance has NIRCAM but with a higher spectral resolution appropriate for a finer characterization of brightest objects.

3.3. Extremely Large Telescopes (>2017)

The ELTs instrumentation for extrasolar planet direct imaging is more prospective, but still it is considered as one of the first priority. Two spectral regimes are considered :

- EPICS (Exoplanet Imaging Camera and Spectrograph for the European ELT, Kasper et al. 2008) and PFI (Planet Finder Imager of the Thirty Meter Telescope, Vasisht et al. 2006) will provide near IR high contrast capability with unprecedented angular resolution (about 10 mas). These two instruments aim at contrasts of $10^8 - 10^9$ as close as 30mas to characterize mature EGPs and possibly large telluric ones.

- METIS (Mid-infrared ELT Imager and Spectrograph, Brandl et al. 2008) has an interesting opportunity to image irradiated planets in the mid IR (3-20μm) down to a few masses of Jupiter (again thanks to the angular resolution gain).

4. Future Projects

In contrary to section 3 the following projects are not yet approved neither planned but many conceptual and technical activities are being pursued. With respect to previous projects, the objectives are focused towards lower masses (telluric planets), older systems (>100 Myr), more distant stars (>100 pc), closer planets (<5 AU) and also shorter wavelengths (<1μm). It is not the goal of this paper to describe all concepts. However, they can be categorized in two main families.

The former Terrestrial Planet Finder Coronagraph was classified as a flag-ship mission. This concept has been re-baselined as a probe-class mission after 2006 with less ambitious objectives. A series of projects like PECO (Guyon et al. 2008), ACCESS (Trauger et al. 2008), EPIC (Clampin 2007) are being studied in the US for the decadal survey. These projects assume a small (1.5-2m) telescope optimized for high-contrast imaging which necessarily means off-axis and good optical quality. Different sort of coronagraphs or wavefront correction devices are proposed to achieve $\sim 10^9$ contrast. The focal instrumentation should be capable of low resolution spectroscopy (R=20-50) in the 0.4-1.0μm range. In Europe, an equivalent mission, the Super Earth Explorer (SEE-COAST, Schneider et al. 2008) was submitted to Cosmic Vision in 2007 (but not selected) with polarimetric capabilities in addition to spectroscopy (see Baudoz et al. 2009, this proceeding). It was ranked as a Medium class mission according to ESA nomenclature which is equivalent to a Probe-class. All these concepts rely for the target sample on planets detected by radial velocity surveys of which the sensitivity extends now to long periods. The prime objective is to achieve a census of giant planets on large

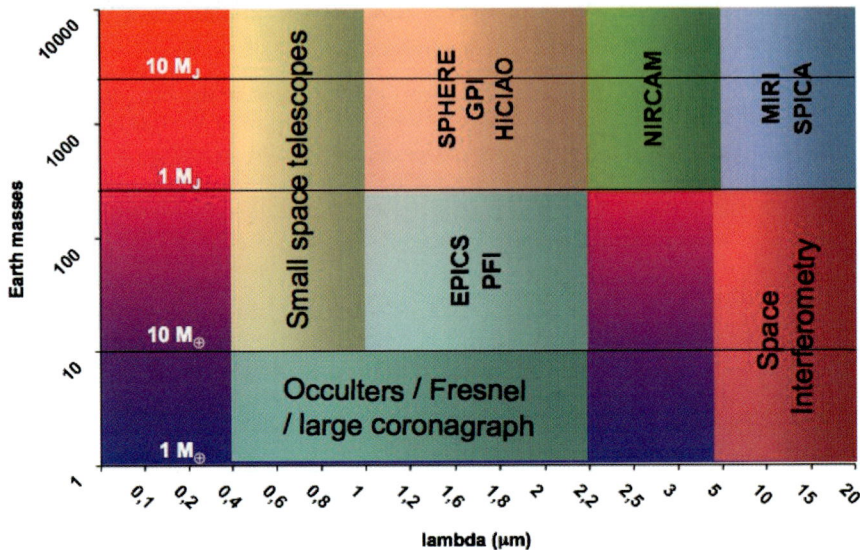

Figure 1. Timeline of direct imaging programs with a separation between ground (bottom) and space (top). Missions names, spectral range and typical detectable planets are indicated in different shadings (blue, green, and red). As a result, the effort should be directed towards mature giant and telluric planets both in the visible and mid IR in order to complement ELTs.

orbits for nearby stars (<20-25pc) and perform spectral characterization. For the nearest stars and the closest planets accessible, large telluric planets could be observed and characterized. A high contrast imaging telescope is also suitable for the identification of circumstellar disks that are intrinsically of interest to study planetary formation and also could eventually prevent the detection of Earth-like planets with even more ambitious missions. The bottom line is to explore the diversity of planets on a pre-defined target sample. Also, it is recognized that the mission cost is driven by the telescope size rather than the instrument and that no major technological change would be needed for bigger telescopes.

A second type of concept has recently emerged. The basic idea is to take advantage of Fresnel diffraction to reach higher contrast. One solution is to fly a large occulter in front of a telescope. A large shadow is projected at the telescope location. Performance are much less sensitive to the telescope wavefront quality but the constraints on the occulter manufacturing can be severe (of the order of 1-10mm precision). The New World Observer proposal (Cash et al. 2008) is intended to detect telluric planets and perform characterization of narrow spectral features in the visible range ideally to pick-up biosignatures. Therefore, it requires a larger telescope than the aforesaid projects. Currently, NWO is made of a 50 m occulter located 80000 km in front of a 4 m telescope. The complexity of formation fly and deployment of large structures in space are some issues that need to be further investigated. An other design is proposed by Koechlin et al. (2005) in a form of an Fresnel imaging lens which is supported by a spacecraft and focalizes the first diffraction order on a field telescope some kilometers away. The lens itself is an array of sub-apertures arranged in a particular form and is uses as a focuser. The image is therefore free of aberrations and could benefit to many coronagraph designs. Combined with pupil apodization, sharp PSFs can be obtained. The design is however chromatic and requires some corrective elements downstream. Several spectral channels would be required to achieve a large bandwidth. The NWO and Fresnel imager are more prospective concepts that need to be elaborated further in the context of space engineering.

It is very likely that none of this concept will actually fly before 2020.

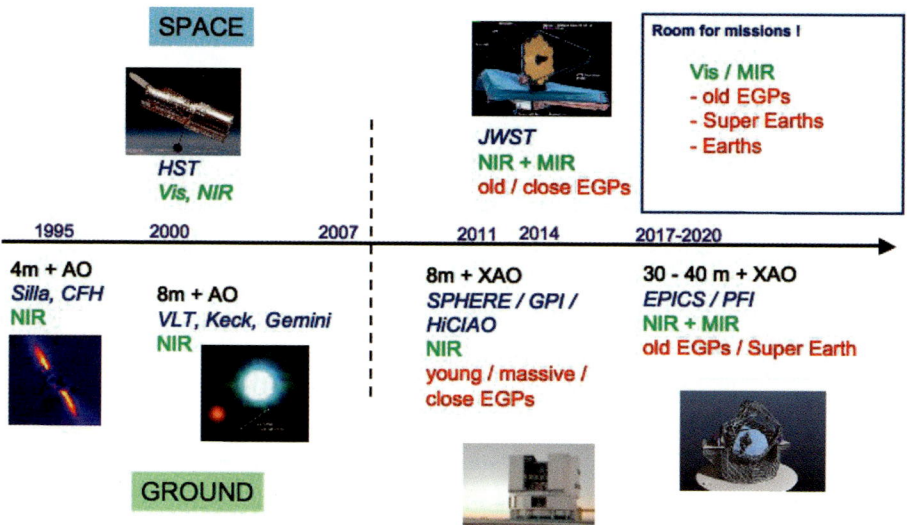

Figure 2. Diagram showing the sweetspot of each mission type as a function of wavelength.

5. Tentative Conclusions

From the above description, it is instructive to put on a timeline the different planned projects identifying spectral range and typical detectable planets. It is obviously recognized that all bandwidths are of interest for the characterization of exoplanets. For the direct imaging, this applies to visible, near IR and mid IR. What comes up from Fig. 1 is that a slot for a space mission geared toward mature giant planets either at visible or mid IR is not covered on the ground at least until 2025-2030. Another way of putting this information in a mass vs. wavelength diagram is shown in Fig. 2. If we apply a concept of minimal effort, this slot could be fill in by small telescopes optimized for a deep characterization of giant planets at visible an ideally down to massive telluric planets that are named Super Earths. If all telluric planets share similar characteristics (atmospheric composition, geophysical attributes, biosignatures, seasonnal activities, ...), it would be worth to devote a space mission on that topic because it would be both complementary to ground-based projects and also preparatory to those that are focused to the search of life on Earth analogs (Cockell et al. 2008).

References

Baudoz, P., Boccaletti, A., Baudrand, J., & Rouan D. 2006, IAU Colloq. 200: Direct Imaging of Exoplanets: Science and Techniques ed. Aime, C. and Vakili, F., pp 553–558.
Baudoz, P., Boccaletti, A., & Rouan, D.J. 2007, Conf. Proc., In the Spirit of Bernard Lyot: The Direct Detection of Planets and Circumstellar Disks in the 21st Century, ed. Kalas P.I., p 258
Baudoz, P., Schneider, J., Boccaletti, A., Galicher, R., Stam, D., & Tinetti, G. (2009) these proceedings.
Beuzit, J.-L., Feldt, M., Dohlen, K., Mouillet, D., Puget, P., Antici, J., Baruffolo, A., Baudoz, P., Berton, A., Boccaletti, A., Carbillet, M., Charton, J., Claudi, R., Downing, M., Feautrier, P., Fedrigo, E., Fusco, T., Gratton, R., Hubin, N., Kasper, M., Langlois, M., Moutou, C., Mugnier, L., Pragt, J., Rabou, P., Saisse, M., Schmid, H. M., Stadler, E., Turrato, M., Udry, S., Waters, R., & Wildi, F. (2006) The Messenger, 125, 29 (ESO)
Boccaletti, A., Riaud, P., Baudoz, P., Baudrand, J., Gratadour, D., Lacombe, F., & Lagrange A.-M. 2004, PASP, 116, 1061
Boccaletti, A., Baudoz, P., Baudrand, J., Reess, J. M., & Rouan D. 2005, Advances in Space Research, 36, 1099
Boccaletti, A., Carbillet, M., Fusco, T., Mouillet, D., Langlois, M., Moutou, C., Dohlen, K. 2008, SPIE, 7105, 1778
Bordé, P. J. & Traub, W. A. 2006, AJ, 638, 488, arXiv:astro-ph/0510597
Brandl, B. R., Lenzen, R., Pantin, E., Glasse, A., Blommaert, J., Venema, L., Molster, F., Siebenmorgen, R., Boehnhardt, H., van Dishoeck, E., van der Werf, P., Henning, T., Brandner, W., Lagage, P.-O., Moore, T. J. T., Baes, M., Waelkens, C., Wright, C., Käufl, H. U., Kendrew, S., Stuik, R., & Jolissaint, L. 2008, SPIE 7014, 558
Burrows, A., Marley, M., Hubbard, W. B., Lunine, J. I., Guillot, T., Saumon, D., Freedman, R., Sudarsky, D., Sharp, C. 1997, Ap.J., 491, 856
Cash, W., Oakley, P., Turnbull, M., Glassman, T., Lo, A., Polidan, R., Kilston, S., & Noecker, C. 2008, SPIE, 7010, 53
Chabrier, G., Baraffe, I., Allard, F., Hauschildt, P. 2000, AJ, 542, 464
Chauvin, G., Lagrange, A.-M., Dumas, C., Zuckerman, B., Mouillet, D., Song, I., Beuzit, J.-L., & Lowrance, P. 2005, A&A, 438, L25
Chauvin, G., Lagrange, A.-M., Zuckerman, B., Dumas, C., Mouillet, D., Song, I., Beuzit, J.-L., Lowrance, P., & Bessell, M. S. 2005, A&A, 438, L29
Clampin, M.J. 2007, Conf. Proc., In the Spirit of Bernard Lyot: The Direct Detection of Planets and Circumstellar Disks in the 21st Century, ed. Kalas, P.I., p 37
Cockell, C. S., Herbst, T., Léger, A., Absil, O. Beichman, C., Benz, W., Brack, A., Chazelas, B., Chelli, A., Cottin, H., Coudé Du Foresto, V., Danchi, W., Defrère, D., den Herder, J.-W., Eiroa, C., Fridlund, M., Henning, T., Johnston, K., Kaltenegger, L., Labadie, L., Lammer, H., Launhardt, R., Lawson, P., Lay, O. P., Liseau, R., Martin, S. R., Mawet, D., Mourard, D., Moutou, C., Mugnier, L., Paresce, F., Quirrenbach, A., Rabbia, Y., Rottgering, H. J. A., Rouan, D., Santos, N., Selsis, F., Serabyn, E., Westall, F., White, G., Ollivier, M., & Bordé, P. 2008, Experimental Astronomy, Sep., 46
Codona, J. L. & Angel, R. 2004 AJ, 604, L117
Doyon, R., Rowlands, N., Hutchings, J., Evans, C. E., Greenberg, E., Scott, A. D., Touhari, D., Beaulieu, M., Abraham, R., Ferrarese, L., Fullerton, A. W., Jayawardhana, R., Johnston, D., Meyer, M. R., Pipher, J., & Sawicki, M., 2008, SPIE 7010, 30
Galicher, R., Baudoz, P., & Rousset, G. 2008, A&A, 488, L9 arXiv:astro-ph:0807.2467
Gay, J., & Rabbia, Y. 1996, Academie des Science Paris Comptes Rendus Serie B Sciences Physiques, 322, 265
Give'on, A., Kasdin, N. J., Vanderbei, R. J., & Avitzour, Y. 2008, SPIE 5905, 368
Guyon, O. 2004, ApJ, 615, 562
Guyon, O., Pluzhnik, E. A., Kuchner, M. J., Collins, B., Ridgway, S. T. 2006, ApJS, 167, 81. arXiv:astro-ph/0608506

Guyon, O., Angel, J. R. P., Backman, D., Belikov, R., Gavel, D., Giveon, A., Greene, T., Kasdin, J., Kasting, J., Levine, M., Marley, M., Meyer, M., Schneider, G., Serabyn, G., Shaklan, S., Shao, M., Tamura, M., Tenerelli, D., Traub, W., Trauger, J., Vanderbei, R., Woodruff, R. A., Woolf, N. J., Wynn, J. 2008, SPIE, 7010, 59
Itoh, Y., Hayashi, M., Tamura, M., Tsuji, T., Oasa, Y., Fukagawa, M., Hayashi, S. S., Naoi, T., Ishii, M., Mayama, S., Morino, J.-I., Yamashita, T., Pyo, T.-S., Nishikawa, T., Usuda, T., Murakawa, K., Suto, H., Oya, S., Takato, N., Ando, H., Miyama, S. M., Kobayashi, N., Kaifu, N. 2005, ApJ, 620, 984 arXiv:astro-ph/0411177
Kalas, P., Graham, J. R., Chiang, E., Fitzgerald, M. P., Clampin, M., Kite, E. S., Stapelfeldt, K., Marois, C., & Krist, J. 2008 ArXiv:astro-ph/0811.1994
Kasper, M. E., Beuzit, J.-L., Verinaud, C., Yaitskova, N., Baudoz, P., Boccaletti, A., Gratton, R. G., Hubin, N., Kerber, F., Roelfsema, R., Schmid, H. M., Thatte, N. A., Dohlen, K., Feldt, M., Venema, L., & Wolf, S. 2008 SPIE 7015, 46
Koechlin, L., Serre, D., Duchon, P. 2005, A&A, 443, 709
Kuhn, J. R., Potter, D., & Parise, B. 2001 ApJ, 553, L189. arXiv:astro-ph/0105239
Krist, J. 2007, Conf. Proc., In the Spirit of Bernard Lyot: The Direct Detection of Planets and Circumstellar Disks in the 21st Century, ed. Kalas P.I., p 49
Kuchner, M. J., & Traub, W. A. 2002, ApJ, 570, 900. arXiv:astro-ph/0203455
Lafrenière, D., Jayawardhana,, R. & van Kerkwijk, M. H. 2008, ApJ689, L153. arXiv:astro-ph/0809.1424
Lagrange, A., Gratadour, D., Chauvin, G., Fusco, T., Ehrenreich, D., Mouillet, D., Rousset, G., Rouan, D., Allard, F., Gendron, E., Charton, J., Mugnier, L., Rabou, P., Montri, J., & Lacombe, F. 2008, arXiv:astro-ph/0811.3583
Lenzen, R., Close, L., Brandner, W., Biller, B., & Hartung, M., 2004, SPIE, 5492, 970
Luhman, K. L. Wilson, J. C. Brandner, W. Skrutskie, M. F. Nelson, M. J. Smith, J. D. Peterson, D. E. Cushing, M. C. Young, E. 2006, ApJ, 649, 894. arXiv:astro-ph/0609187
Luhman, K. L., Patten, B. M., Marengo, M., Schuster, M. T., Hora, J. L., Ellis, R. G., Stauffer, J. R., Sonnett, S. M., Winston, E., Gutermuth, R. A., Megeath, S. T., Backman, D. E., Henry, T. J., Werner, M. W., Fazio, G. G., 2007, ApJ, 654, 570. arXiv:astro-ph/0609464
Lyot, B. 1939, MNRAS, 99, 538
Macintosh, B., Graham, J., Palmer, D., Doyon, R., Gavel, D., Larkin, J., Oppenheimer, B., Saddlemyer, L., Wallace, J. K., Bauman, B., Evans, J., Erikson, D., Morzinski, K., Phillion, D., Poyneer, L., Sivaramakrishnan, A., Soummer, R., Thibault, S., Veran, J.-P. 2006, SPIE 6272, p11
Marois, C., Lafrenière, D., Doyon, R., Macintosh, B., & Nadeau, D. 2006 ApJ, 641 556
Marois, C., Macintosh, B., Barman, T., Zuckerman, B., Song, I., Patience, J., Lafrenière, D., & Doyon, R. 2008 Science 322, 1348. arXiv:astro-ph/0811.2606
Marois, C., Macintosh, B., Soummer, R., Poyneer, L., & Bauman, B. 2008, SPIE, 7015, 47
Neuhäuser, R., Guenther, E. W., Wuchterl, G., Mugrauer, M., Bedalov, A., Hauschildt, P. H. 2005, A&A, 435, L13
Quirrenbach, A. 2005, arXiv:astro-ph/0502254
Racine, R., Walker, G. A. H., Nadeau, D., Doyon, R., Marois, C. 1999, PASP, 111, 587
Rouan, D., Riaud, P., Boccaletti, A., Clénet, Y., Labeyrie, A. 2000, PASP, 112, 1479
Schmid, H. M.,Beuzit, J.-L., Feldt, M., Gisler, D., Gratton, R., Henning, T., Joos, F., Kasper, M., Lenzen, R., Mouillet, D., Moutou, C., Quirrenbach, A., Stam, D. M., Thalmann, C., Tinbergen, J., Verinaud, C., Waters, R., Wolstencroft, R. 2006, IAU Colloq. 200: Direct Imaging of Exoplanets: Science and Techniques, ed. Aime, C., & Vakili, F., pp 165-170
Schneider, J. Boccaletti, A. Mawet, D. Baudoz, P. Beuzit, J. L. Doyon, R. Marley, M. Stam, D. Tinetti, G. Traub, W. Trauger, J. Aylward, A. Cho, J. Y. K. Keller, C. U. Udry, S. & for the SEE-COAST Team, 2008, ArXiv:astro-ph/0811.3908
Sparks, W. B., & Ford, H. C. 2002, ApJ, 578, 543. arXiv:astro-ph/0209078
Thalmann, C., Schmid, H. M., Boccaletti, A., Mouillet, D., Dohlen, K., Roelfsema, R., Carbillet, M., Gisler, D., Beuzit, J.-L., Feldt, M., Gratton, R., Joos, F., Keller, C. U., Kragt, J., Pragt, J. H., Puget, P., Rigal, F., Snik, F., Waters, R., Wildi, F. 2008, SPIE, 7014, 112

Tamura, M., Hodapp, K., Takami, H., Abe, L., Suto, H., Guyon, O., Jacobson, S., Kandori, R., Morino, J., -I., Murakami, N., Stahlberger, V., Suzuki, R., Tavrov, A., Yamada, H., Nishikawa, J., Ukita, N., Hashimoto, J., Izumiura, H., Hayashi, M., Nakajima, T., & Nishimura, T. 2006, SPIE, 6269, 28

Trauger, J. T., & Traub, W. A. 2007, Nat, 446. 771

Trauger, J., Stapelfeldt, K., Traub, W., Henry, C., Krist, J., Mawet, D., Moody, D., Park, P., Pueyo, L., Serabyn, E., Shaklan, S., Guyon, O., Kasdin, J., Spergel, D., Vanderbei, R., Belikov, R., Marcy, G., Brown, R. A., Schneider, J., Woodgate, B., Matthews, G., Egerman, R., Polidan, R., Lillie, C., Ealey, M., Price, T. 2008, SPIE 7010, 69

Vasisht, G., Crossfield, I. J., Dumont, P. J., Levine, B. M., Troy, M., Shao, M., Shelton, J. C., & Wallace, J. K. 2006, SPIE 6272, 161

Vigan, A., Langlois, M., Moutou, C., & Dohlen, K. 2008 ApJ, 489, 1345. arXiv:astro-ph/0808.3817

The Potential of High Contrast Coronagraphy

E. Serabyn

Jet Propulsion Laboratory, California Institute of Technology, 4800 Oak Grove Drive., Pasadena, CA 91109, USA

Abstract. The direct detection of faint companions near much brighter stars requires the development of very high-contrast, small field-of-view detection techniques, and the past decade has seen remarkable conceptual and instrumental progress in this area. New coronagraphic techniques are being developed and deployed, as are extreme adaptive optics (ExAO) systems that will enable the advantageous exploitation of these new techniques. This paper provides a short overview of promising high contrast coronagraphic techniques, as well as recent examples of ExAO coronagraphy and transit measurements obtained with the ExAO-level "well-corrected subaperture" at Palomar.

1. Introduction

As is discussed in this volume, rapid progress is being made in the search for planets around nearby stars. Radial velocity (RV) measurements have led to the detection of hundreds of exoplanets, and transit measurements are beginning to add significantly to the tally. Indeed, transit measurements have also enabled spectroscopic measurements of select exoplanets. While these techniques will no doubt provide ever increasing observational data, a technique which is only recently starting to come into its own is the direct imaging of exoplanets. Indeed, the recent companion detections discussed in this volume highlight both the promise and the difficulty of direct imaging. The ultimate goal of course is the ability to directly image nearby solar systems, so as to allow both direct observations of orbital motions and phases, as well as spectroscopic measuremements of each of the exoplanets. Direct imaging has a further advantage in that it can potentially access a region of parameter space complementary to those accessible to the RV and transit techniques, since it is particularly sensitive to companions at large separation, i.e., those with long orbital periods.

The main obstacles to the direct imaging of nearby exoplanets are their very low brightness ratios (or contrasts) relative to their host stars, combined with their small separations. In reflected light, an exact terrestrial planetary analog located 1 AU from a G2 star would be only slightly more than 10^{-10} as bright as the star. Because of its larger diameter, a jovian planet at the same orbital radius would be roughly a factor of 100 brighter. Since reflected light decreases as r^{-2}, a measurement capability at either a smaller inner working angle (IWA), or a deeper contrast, c, would be beneficial for detecting exoplanets in reflected light, as either would expand the size of the accessible search space (inward and downward, respectively). Note that because of this r^{-2} brightness dependence, the IWA and attainable contrast, c, can be traded off against each other, i.e., a system with a smaller IWA can have a more modest contrast capa-

Table 1. Coronagraphic Alternatives

	Pupil Plane	Focal Plane
Amplitude:	Amplitude apodization Shaped pupil Pupil remapping External Occulter	Hard-edged mask Graded-intensity mask Band-limited mask
Phase:	Pupil phase apodization	Circular phase mask Four quadrant phase mask Scalar vortex mask Vector vortex mask

bility. One could thus combine these two quantities into a figure of merit reflecting the rough size and depth of the (inner edge of the) accessible search area. For example, one could define a "coronagraphic area", as $A = c * IWA^2$ (with the units of angular area). The reason for such a definition is that the A needed for detection of a given type of exoplanet at a given distance from us is a constant, independent of the star-planet separation. Moreover, A also gives the maximum distance to which such an exoplanet can be detected with a given coronagraph.

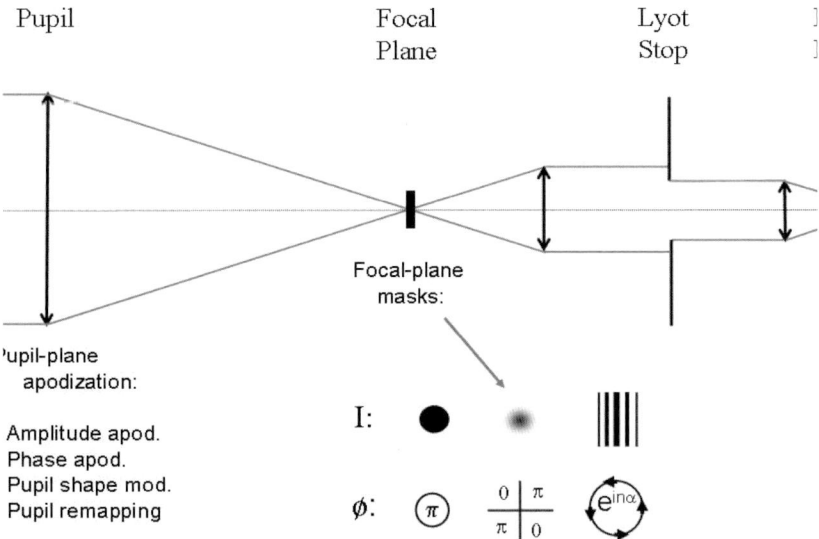

Figure 1. Schematic coronagraph layout, with pupil and focal plane modifications indicated. Possible focal plane masks include: First row: hard edged blocker, graded attenuator and band-limited mask. Second row: circular phase mask, four-quadrant phase mask and vortex mask.

Young jovian planets also contribute their own emitted light in the near-IR, which is independent of distance from the star, allowing for much easier detection possibilities, potentially to about the 10^{-4} level. At the current time, attainable contrasts at very small angles are limited to $\sim 10^{-4}$ or so also, enough to motivate initial searches with existing capabilities. However, both coronagraphic and wavefront sensing and control techniques have been undergoing significant advances recently, leading to expectations of significantly improved performance with next generation ExAO systems to come on line in the near future. This paper summarizes promising techniques for high-contrast coronagraphy, focusing on physical principles, and concludes with recent examples of initial coronagraphy and exoplanet transit measurements made with an existing ExAO system. Space limitations prevent addressing fully all of the exciting developments in these fields.

2. High Contrast Coronagraphy

Coronagraphy began with the solar coronagraph of Lyot (1938), which blocked the sunlight collected by a single telescope aperture while allowing transmission of light from the off-axis coronal light. In the intervening years, and especially in the past decade, a number of quite novel and promising alternatives for stellar coronagraphs have emerged. Coronagraphic approaches to starlight suppression still rely on the modification of the incoming stellar electric field distribution prior to the generation of the final focal plane image, but potential modifications may be applied either in a pupil plane or a focal plane, and either to the amplitude or the phase of the field. A number of potential coronagraphic alternatives are summarized in Table 1, based on this fourfold characterization. For reference, a conceptual layout of a general coronagraph is shown in Fig. 1. Guyon et al. (2006) compared the imaging properties of the coronagraph alternatives known at the time, and based on the metric of "useful throughput", found that pupil remapping and vortex phase mask coronagraphy (see below) are closest to ideal coronagraphic performance. Several of the promising coronagraphic alternatives are now briefly discussed.

A classical amplitude-based Lyot coronagraph rejects starlight using two blockers. First, a small opaque mask several λ/D in diameter is centered on the focal plane stellar image, in order to block the innermost part of the point spread function (PSF). Only the high frequency components of the PSF then propagate to a subsequent pupil plane, where they appear localized near the pupil periphery. An opaque undersized pupil stop (the Lyot stop) is then used to further reduce the residual starlight (Fig. 1; see also Sivaramakrishnan et al. 2001).

Because of the Fourier transform relationship between the focal plane and the pupil plane, a hard-edged focal plane stop wil lead to ringing in the pupil plane, so that the light distribution in the subsequent Lyot plane will not be sharply confined near the edge of the pupil. There will thus be significant leakage of starlight into the pupil. Luckily, this ringing can be decreased by using a focal plane stop with a tapered absorption. In the ideal band-limited case (Kuchner & Traub 2002), very high stellar rejection can be provided by the Lyot stop. Fig. 2 shows an example of the very well-defined ring-like pupil plane distribution recently measured in our lab for a prototype band-limited mask intended for possible use in the JWST NIRCAM (Krist et al. 2007). In the laboratory, focal plane coronagraphy with band-limited masks have demonstrated narrowband contrasts of several 10^{-10} (Trauger and Traub 2007), close to that required for terrestrial exoplanet detection, and work on broadening the passband is also proceeding.

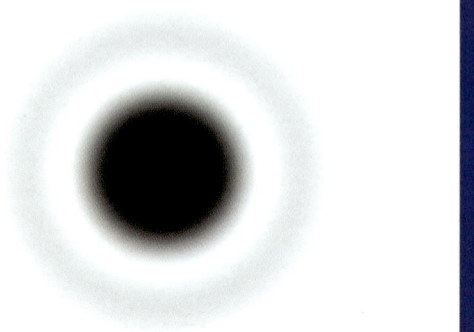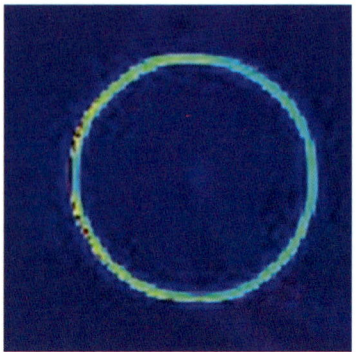

Figure 2. Measured ring-like pupil (right) generated with a band-limited mask (left) manufactured at JPL, as per Krist et al. (2007). An undersized Lyot stop can then easily reject the light in the ring-like pupil.

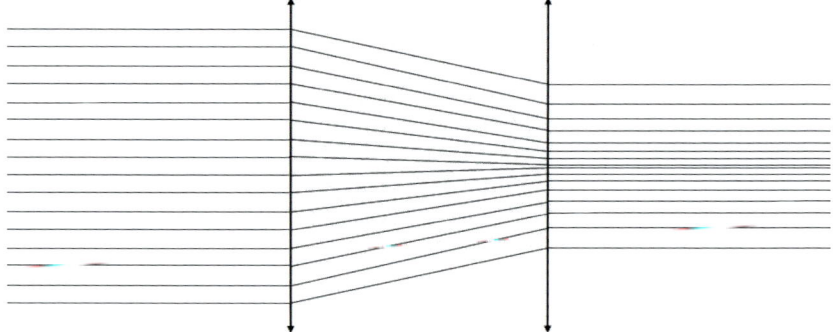

Figure 3. Schematic illustrating the redistribution of light in the pupil in the PIAA approach, described in Guyon (2003).

On the other hand, the pupil plane amplitude distribution can also be modified, as in classical apodization. The goal is then to smoothly taper the pupil plane field, leading, after Fourier transformation, to a focal plane spot well-localized near the optical axis, so that little starlight leaks to large angles. One way to effect such an apodization is with a spatially variable pupil plane attenuator. Optimized apodization solutions have been found (e.g. Kasdin et al. 2003), aiming at e.g., the smallest inner working angle (IWA), or at generating a deep "dark hole" search region near the optical axis (Gonzalves & Nisenson 2003). Note that most of the high-contrast coronagraphs planned to come on line on the ground and in space (e.g. JWST) in the next few years include a variation of a classical amplitude-based Lyot coronagraph.

Alternatively, the shape of the pupil can also be modified, so as to provide a focal plane image with localized dark search regions. The simplest pupil shape modification is a square aperture, which is known to have low diffraction along the PSF diagonals (Nisenson & Papaliolios 2001), and more complex pupil shapes also can scatter light preferentially in certain directions or areas, leaving other search regions dark (Kasdin et al. 2003).

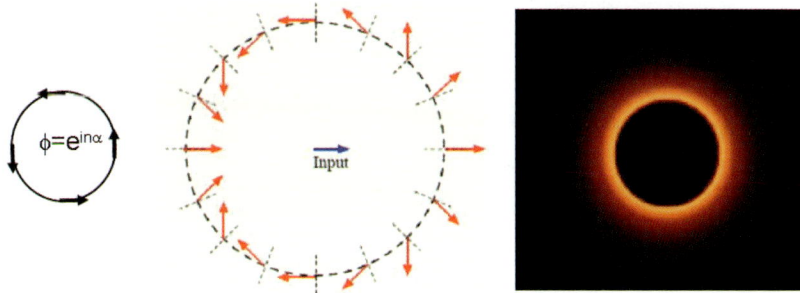

Figure 4.   Illustration of the operation of vector vortex coronagraph (Mawet et al. 2008), which uses an azimuthally varying halfwave plate to induce an azimuthal phase variation (left) in the focal plane. The optical axis rotates with azimuth, and the output polarization rotates twice as fast (center). The fast axis orientations are shown as light dashed lines, while the output field orientation, for the input polarization shown, is given by the arrows. Right: the ideal pupil plane distribution after passage through a vortex mask.

However, while intensity apodization does reduce the PSF wings, this generally comes at the cost of lower throughput, because of the loss of pupil area. On the other hand, beam shaping can also be used to alter the transverse beam profile. In the context of coronagraphy, this is referred to as "phase-induced amplitude apodization" (PIAA) (Guyon 2003) or pupil remapping (Traub & Vanderbei 2003). In effect, the pupil is laterally remapped (distorted) to provide the desired tapered distribution. Conceptually, by spreading the outer rays further and further apart (Fig. 3), the incident flat-topped beam is converted to a tapered beam profile (Guyon 2003). Because the PIAA does not rely on attenuation to produce the tapered distribution, it is potentially highly efficient, and can be used to observe quite close to the axis.

For space-based applications, one could also block the starlight by means of an external occulter located well upstream of the telescope (Cash 2006). The occulter must be far enough in front of the telescope to allow planetary light arriving at small off-axis angles to pass by unobstructed, which leads to very large reconfiguration motions when changing targets.

On the other hand, the phase of the field can also be modified, and three potential focal plane modifications are illustrated in Fig. 1. The four-quadrant phase mask coronagraph applies alternating phase shifts of π radians in the four quadrants of the focal plane, and so avoids lateral size-related chromaticities (Rouan et al. 2000). This technique has been applied at the VLT (e.g. Boccaletti et al. 2008) and at Palomar (Haguenauer et al. 2006) and will be included in the JWST mid-infrared instrument. Indeed, brown dwarfs as close to the optical axis as 2.8 λ/D have now been detected with phase-mask coronagraphy (Serabyn et al. 2009).

A more recent development is the vortex phase-mask coronagraph, in which an azimuthal phase ramp is applied to the focal plane field. The phase ramp can be generated either by a spatially variable glass thickness (Foo, Palacios & Swartzlander 2005),

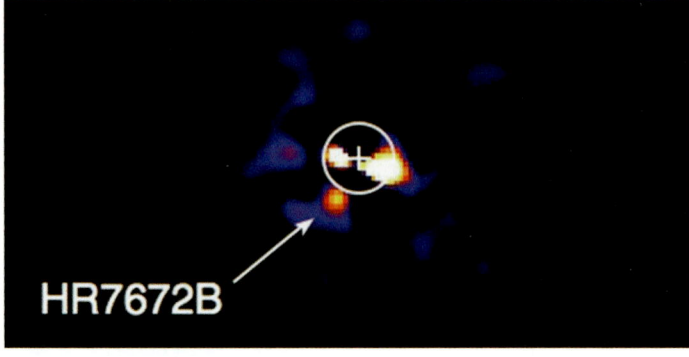

Figure 5. Image of the brown dwarf companion HR7672B obtained with the 1.6 m Palomar WCS. The circle has a radius of 2 λ/D.

or polarization-rotation leading to variations in the Pancharatnam phase (Fig. 4; Mawet et al. 2005). The latter approach has maximum search space potential because it has no phase discontinuities except along the optical axis. Such phase-based coronagraphs yield pupil plane distributions in the Lyot plane wherein the starlight lies beyond the original pupil (e.g. Fig. 4). An opaque undersized Lyot stop can then be very effective in removing starlight. Furthermore, because there is no opaque mask present at the center of the focal plane, very small inner working angles can be achieved with phase mask coronagraphy.

On the other hand, the phase of the pupil plane field can also be modified. It is well-known that pupil-plane phase aberrations modify the focal plane distribution. A theoretical treatment of pupil plane phase modification has been provided by Yang et al. (2004), and initial on-sky tests and demonstrations have recently been made (Codona et al. 2006; Serabyn et al. 2007).

3. Recent Progress and Examples

Although ExAO coronagraphic systems do not yet exist on large telescopes, there is one ExAO test facility available already - the ExAO-level "well-corrected subaperture" (WCS) on the Palomar 200-inch telescope (Serabyn et al. 2007). To make up for the WCS's small aperture diameter, a phase mask coronagraph with a small IWA has been employed with it. This system was recently used to image known brown dwarf companions (Serabyn et al. 2009), and Fig. 5 shows an example of a brown dwarf clearly seen at an offset of only 2.8 λ/D. As this image was obtained with an aperture of only 1.6 m, the prospects for future high-contrast imaging on large telescopes equipped with ExAO seem very promising indeed.

Finally, the stability of the PSF provided by an ExAO system brings additional benefits. In particular, the advent of ExaO will also enable a new method of making transit observations from the ground. Fig. 6 shows an example of the first half of a primary transit of HD189733 recently observed with the Palomar WCS. The flux decrease of a couple of percent at the initiation of the transit is clearly already seen in the raw data. This again indicates a very promising future when ExAO becomes available on large telescopes.

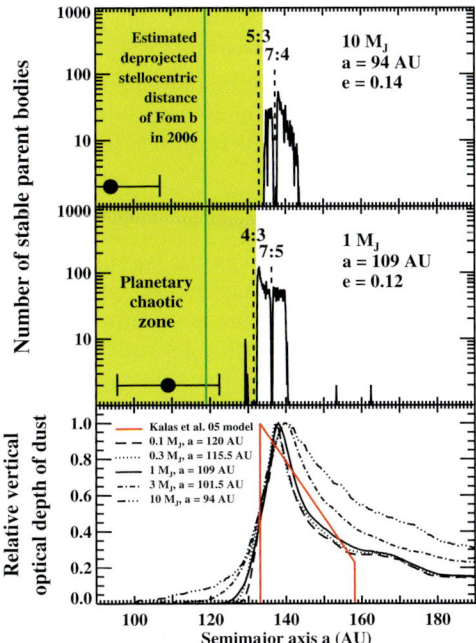

Figure 3. Dynamical models of how Fomalhaut b gravitationally sculpts the belt from Chiang et al. (2009). (Top and middle) Histograms of time averaged semimajor axes of parent bodies that survive 100-My integrations with Fomalhaut b, whose parameters are chosen to reproduce the beltÕs inner edge at 133 AU and ellipticity of 0.11. Parent bodies are evacuated from Fomalhaut bÕs chaotic zone (shaded region). Gaps open at the planetÕs resonances, akin to the solar systemÕs Kirkwood gaps. Black circles and bars mark the range of stellocentric distances spanned by the model orbits for Fomalhaut b. The apocentric distance for 10 MJ is inconsistent with the observed stellocentric distance of Fomalhaut b (green line). The 1-MJ model is consistent. (Bottom) Vertical optical depth profiles of dust generated from parent bodies. The planet orbit is tuned so that the optical depth is at half maximum at 133 AU, the location of the inner edge of the scattered-light model from ?, which itself is an idealized and non-unique fit to the HST data. Although the dynamical and scattered-light models do not agree perfectly, lower planet masses are still inferred because they do not produce broad tails of emission at a >~140 AU. At a >~ 160 AU, the HST data are too uncertain to constrain any model.

3. Constraints on Planetary properties

The presence of a resolved dust belt in the Fomalhaut system offers the opportunity to independently constrain the mass of Fomalhaut b via its interaction with the dust belt. (Chiang et al. 2009) have conducted an extensive analysis of the Fomalhaut system by modeling gravitational influence of Fomalhaut b on the dust belt, to reproducing the dust-belt properties inferred from the HST scattered-light images. Our model assumes that Fomalhaut b is solely responsible for the observed belt morphology. This assumption implies that the orbits of the belt and of Fomalhaut b are apsidally aligned. The deprojected space velocity and current location of Fomalhaut b are nominally inconsistent with this expectation. Apsidal misalignment may imply the existence of additional perturbers; then the mass estimates derived from our single-planet models are upper limits.

Our modeling procedure comprises four steps and is discussed in detail by (Chiang et al. 2009). The first step involves creating a population of several thousand parent bodies stable to gravitational perturbations. Initial parent body orbits have semimajor axes between 120 and 140 AU, and their eccentricities and longitudes of periastron are purely secularly forced by the planet (Murray and Dermott 1999). The forced orbits thus constructed over 100 Myr are nested ellipses with eccentricity of 0.11 that approximate the observed belt morphology. This elliptical annulus of parent bodies is termed a Òbirth ringÓ (Strubbe and Chiang 2006); erosive collisions among parent bodies give birth to smaller but more numerous dust grains, which are the predominant source of the observed scattered stellar light. The second step of our procedure is to track dust trajectories such that each parent releases a dust grain with the same instantaneous position and velocity as its parentÕs. The trajectory of a grain is then integrated forward under the effects of radiation pressure and Poynting-Robertson drag, over a range of βs. Third, we superpose the various β integrations to construct maps of optical depth normal to the belt plane. We compute the optical depth presented weighting each β integration according to a Dohnanyi (Dohnanyi 1969) grain size distribution. The final step is to compare the optical depth profile of our dynamical model with that of a scattered-light model adjusted to fit the 2004 HST image of FomalhautÕs belt (?). We focus on the one belt property that seems most diagnostic of planet mass and orbit: the beltÕs inner edge, having a semimajor axis of a_{inner} = 133 AU according to the scattered-light model.

Two trends that emerge from our modeling imply that the mass of the planet should be low. First, as mass increases, the planet more readily perturbs dust grains onto eccentric orbits, and the resultant optical depth profile becomes too broad at distances greater than ~140 AU. Second, for the belt to remain undisrupted, larger-mass planets must have smaller orbits, violating our estimate for the current stellocentric distance of Fomalhaut b. Together, these considerations imply that $M < 3\ M_J$). This upper limit supersedes those derived previously (Quillen 2006), as the quantitative details of our model are more realistic (Chiang et al. 2009). The single planet dynamical model predicts that planet and belt orbits be apsidally aligned. If future astrometry bears this out, then the dynamical model, coupled with the observed positions of Fomalhaut b, lead to a firm estimate of the mass and orbit of Fomalhaut b: $M = 0.5 M_J$, $a = 115$ AU, and $e = 0.12$ (Chiang et al. 2009).

4. Summary

Fomalhaut b, is the first detection of a gas giant planet in the visible since the discovery of Neptune. The presence of Fomalhaut b was predicted from the morphology of the dust belt, suggesting that resolved imaging of debris disks presents a new avenue for exoplanet searches. Fomalhaut b's association with the dust belt also provides an independent constraint can be applied to determine the mass of the planet. Via the modeling of its gravitational influence on the dust belt, the planetÕs mass is at most $3M_J$. While the planet's photometry matches a T_{eff} 400K model atmosphere at 0.8 μm, at longer wavelengths uncertainties in the model atmospheres complicate the interpretation of our near and mid-IR photometry. The brightness at 0.6 μm and the lack of detection at longer wavelengths suggest that the detected flux may include starlight reflected off a circumplanetary disk, with dimension comparable to the orbits of the Galilean satellites. Further observations of Fomalhaut b are required to constraint its orbital parameters and determine its spectral energy distribution.

Acknowledgments. Supported by HST programs GO-10598 (P.K.) and GO-10539 (K.S. and J.K.), provided by NASA through a grant from the Space Telescope Science Institute (STScI) under NASA contract NAS5-26555; NSF grant AST-0507805 (E.C.). We thank the staff at STScI, Keck, and Gemini for supporting our observations.

References

Ardila, D., D. R., Golimowski, D. A., Krist, J. E., Clampin, M., Williams, J. P., Blakeslee, J. P., Ford, H. C., Hartig, G. F., & Illingworth, G. D. 2004, ApJ 617, L147
Aumann, H. H., Beichman, C. A., Gillett, F. C., de Jong, T., Houck, J. R., Low, F. J., Neugebauer, G., Walker, R. G. & Wesselius, P. R. 1984, ApJ 278, L23
Barrado y Navascues, D. 1998, Astron. Astrophys. 339, 831
Burrows, A., Sudarsky, D., Lunine, J. I., 2003, ApJ 596, 587
Chiang, E., Kite, E., Kalas, P., Graham, J. R. & Clampin, M. 2009, ApJ, 693, 734-749
Dohnanyi, J. W. 1969, J. Geophys. Res. 74, 2531
Fortney, J. J. et al. 2008, AJ 683, 1104
Greaves, J. S., Holland, W. S., Wyatt, M. C., Dent, W. R. F., Robson, E. I., Coulson, I. M., Jenness, T., Moriarty-Schieven, G. H., Davis, G. R., Butner, H. M., Gear, W. K., Dominik, C., Walker, H. J. 2005, ApJ 619, L187
Holland, W. S., Greaves, J. S., Zuckerman, B., Webb, R. A., McCarthy, C., Coulson, I. M.; Walther, D. M., Dent, W. R. F., Gear, Walter K., Robson, I. 1998, Nature, 392, pp. 788-791.
Kalas, P., Graham, J. R. & Clampin, M. 2005, Nature 435, 1067
Kalas, Paul; Graham, James R.; Clampin, Mark C.; Fitzgerald, Michael P. 2006, ApJ 637, L57
Kalas, P., Graham, J. R., Chiang, E., Fitzgerald, M. P., Clampin, M., Kite, E. S., Stapelfeldt, K., Marois, C., & Krist, J. 2008 Science, 322, 1345
Moro-Martin, A. & Malhotra, R. 2002, AJ 124, 2305
Murray, C. D. & Dermott, S. F 1999, Solar System Dynamics, (Cambridge Univ. Press, Cambridge, 1999)
Ozernoy, Leonid M., Gorkavyi, N. N., Mather, John C., & Taidakova, Tanya A. 2000, ApJ 537, L147
Quillen, A. C. 2006, MNRAS 372, L14.
Smith, B. A. & Terrile, R. J. 1984, Science 226, 1421
Strubbe,L. E. & Chiang, E. I. 2006, Astrophys. J. 648, 652
Weinberger, A. J., Becklin, E. E., Schneider, G.; Smith, B. A., Lowrance, P. J., Silverstone, M. D., Zuckerman, B. & Terrile, R. J. 1999, ApJ 525, L53
Wyatt, M. C., Dermott, S. F., Telesco, C. M., Fisher, R. S., Grogan, K., Holmes, E. K. & PiÛa, R. K. 1999, ApJ 527, 918
Wyatt, M. 2005, in Proceedings of the Miniworkshop on Nearby Resolved Debris Disks. October 19-20, 2005. Space Telescope Science Institute, Baltimore, MD, USA. Edited by Inga Kamp, Margaret Meixner, p.42

Keigo Enya representing the SPICA team.

High-Contrast Imaging:
A Wider View on Extrasolar Planetary Systems

M. Bonavita,[1,2] R. U. Claudi,[1] G. Tinetti,[3] J.-L. Beuzit,[4] G. Chauvin,[4] S. Desidera,[1] R. Gratton,[1] and M. Kasper[5]

Abstract. Although very successful (more than 350 planets discovered up to now) indirect methods for extrasolar planet detection (radial velocity, transits) are sensitive to planets quite close to their hosts.
   Moreover, accurate studies of planet characteristics are feasible only for a subset of object which are strongly irradiated. Standing at this point, any information about the exoplanets in wide orbits (more than 5-10 AU) is missing.
   High contrast imaging could be the key to open us a door to an unexplored region of star planet separation and to shed light on these unknown far away worlds.
   But it's not just a matter of detections. In fact coupling integral field spectrographs to extreme adaptive optic modules at the focus of 8m class telescopes (SPHERE for VLT and GPI for South Gemini), and in the future to ELTs (EPICS), would allow us to perform a first order characterization of the exoplanets themselves.
   Here we present the potential of the high contrast imaging technique, comparing it's capabilities with the ones of the indirect methods.

1. Introduction

The radial velocity technique (RV) is currently the most successful method for the detection of extrasolar planetary systems. The searches are mainly focused on quite old stars, which ensure high RV precision due also to the low activity level. On the other hand, the transit technique brought informations on radius and density of giant planets allowing us to probe their internal structure, again focusing on close-in planets.
   Deep imaging observations represent the most viable technique to extend such systematic characterization at larger scales, reaching a complementary set of targets with respect to the indirect techniques, adapted for the characterization of the inner part of exo-planetary systems.

---

[1]INAF - Osservatorio Astronomico di Padova, Vicolo dell'Osservatorio 5, 35122 Padova, Italy

[2]Universitá degli Studi di Padova - Dipartimento di Astronomia, Vicolo dell'Osservatorio 3, 35122 Padova, Italy

[3]University College London, Gower street, London WC1E 6BT, UK

[4]Laboratoire d'Astrophysique de l'Observatoire de Grenoble, UniversitÃl' Joseph Fourier, CNRS, BP 53, 38041 Grenoble, France

[5]European Southern Observatory, Karl-Schwarzschild Strasse 2, D-84748 Garchin-bei-München, Germany

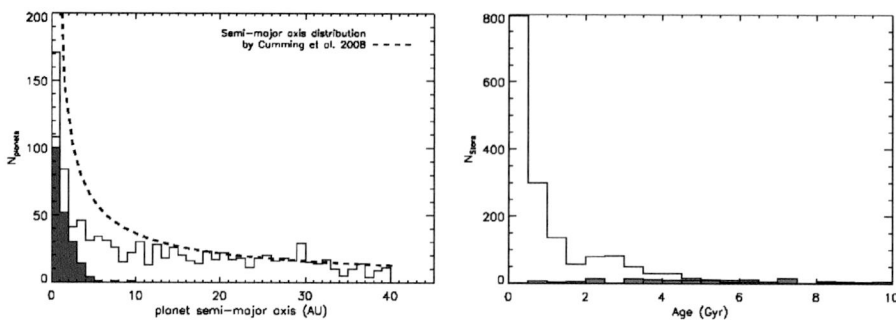

Figure 1. **Left:** Histogram of semi-major axis of planets discovered by RV techniques (shaded area) superimposed on those predicted following the distributions by Cumming et al. (2008), extrapolated up to 20 AU. **Right** Histogram of stellar ages (Gyrs) for the stars with known planets (shaded area) compared to those selected for the preliminary sample optimized for SPHERE and EPICS observations.

2. Laying the Basis for Future Studies

Deep imaging surveys have recently focused on young, nearby associations. Due to their youth and proximity, they offer an ideal niche to detect close planetary mass companions which are hotter and brighter. Various deep imaging surveys have been recently completed using different direct imaging techniques (coronagraphy, differential imaging, L-band imaging). A significant number have reported a null-detection result. This result could however give important constraints on physical and orbital properties (mass, period, eccentricity distributions) of giant planets.

Fig. 2, which refers to the imaging survey of young, nearby associations recently done with NACO (see Chauvin et al. 2008, submitted), shows some example of the statistical analysis that can be done using these informations (see also Kasper et al. 2007; Lafrenière et al. 2007; Nielsen et al. 2008, for similar works).

Despite the model-dependency of the mass prediction, these kind of analysis could be used as a starting point for the searches to be done with the next generation instruments like SPHERE (Beuzit et al. 2008) and GPI (Macintosh et al. 2008), that will show us far away Jupiter-like planets shining by their intrinsic light, focusing on relatively young stars (as showed in right panel of Fig.1), thus being the ideal continuation of the present efforts.

3. Extremely Large Telescopes: the Last Frontier

ELT instruments like EPICS (Kasper et al. 2008) will represent the ideal link between direct and indirect detections. In fact they will allow us to observe young, nearby systems discovered by next generation imagers and also to obtain the first images of planets already detected by RV (as showed in Fig. 3). Moreover, this kind of instruments will also be sensitive at planets shining through their reflected light, thus reaching for the first time the outer zones of planetary systems around stars similar to our Sun. Finally, in some favourable cases, they will be sensitive enough to detect rocky planets in the habitable zones of their hosts.

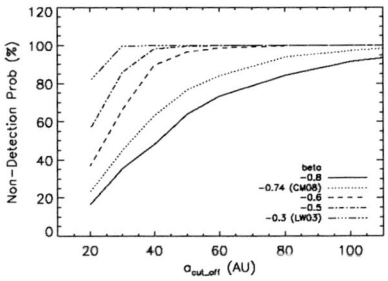 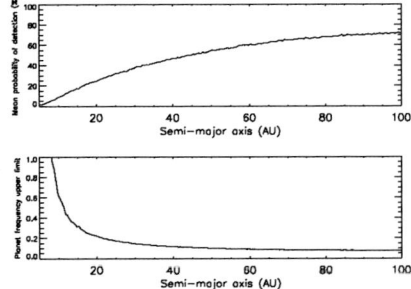

Figure 2. **Left:** Significance of the null result of NACO survey compared with the prediction based on semi-major axis power-law distributions with various index (β), extrapolated at different cut-offs. The vales of β from Cumming et al. (2008) (CM08) and Lineweaver & Grether (2003) (LW03) are considered as reference values. **Right:** Probability of null detection (upper panel) and upper limit on planet frequency (lower panel) based on the result of NACO ASSO survey.

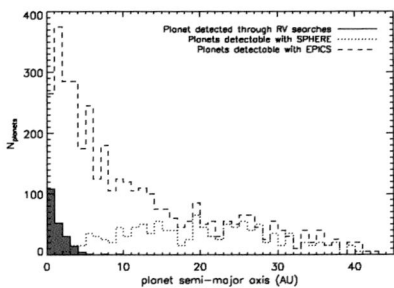 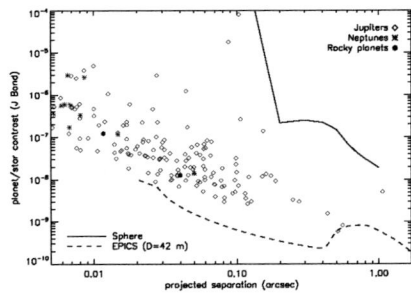

Figure 3. **Left:** Histogram of semi-major axis of planets discovered by RV techniques (shaded area), compared to those detectable with SPHERE (dotted line) and EPICS (dashed line) **Right:** Planet-star contrast in J-Band VS projected separation for the planets already discovered by RV, with superimposed the detection limits for SPHERE (solid line) and EPICS (dashed line). Jupiter-like planets ($M_p > 40 M_{Earth}$), Neptune-like planets ($10 M_{Earth} \leq M_p \leq 40 M_{Earth}$) and Rocky planets ($M_p < 10 M_{Earth}$) are plotted as diamonds, asterisks and filled circle, respectively.

4. Conclusions

As RV measurements, in the last 15 years, have given us informations on close-in giant planets around solar-type stars, direct imaging seems to be able to unveil, in the next decades, a different kind of extrasolar planetary systems. The available measurements are already giving us indirect informations on far away planets around young stars, but only passing trough the intermediate step of next generation imagers and finally with the advantage of ELT instruments we will have a wide view on planetary systems at different stages of their evolution.

References

Beuzit, J.-L., Feldt, M., Dohlen, K., Mouillet, D., Puget, P., Wildi, F., Abe L., Antichi J., Baruffolo, A., Baudoz, P., Boccaletti, A., Carbillet, M., Charton, J., Claudi, R., Downing, M., Fabron, C., Feautrier P., Fedrigo, E., Fusco, T., Gach, J.-L., Gratton, R., Henning, T., Hubin, N., Joos, F., Kasper, M., Langlois, M., Lenzen, R., Moutou, C., Pavlov, A., Petit, C., Pragt, J., Rabou, P., Rigal, F., Roelfsema, R., Rousset, G., Saisse, M., Schmid, H.-M., Stadler, E., Thalmann, C., Turatto, M., Udry, S., Vakili, F.,& Waters, R. 2008, SPIE, 7014, 41

Cumming, A., Butler, R. P., Marcy, G. W., Vogt, S. S., Wright, J. T., & Fischer, D. A. (2008) PASP, 120, 531âĂŞ554

Kasper, M., Apai, D., Janson, M., & Brandner, W. (2007) A&A 472, 321-327

Kasper, M. E. Beuzit, J.-L. Verinaud, C. Yaitskova, N. Baudoz, P. Boccaletti, A. Gratton, R. G. Hubin, N. Kerber, F. Roelfsema, R. Schmid, H. M. Thatte, N. A. Dohlen, K. Feldt, M. Venema, L. & Wolf, S. 2008, SPIE, 7015, 31

Lafrenière, D., Doyon, R., Marois, C., Nadeau, D., Oppenheimer, B. R., Roche, P. F., Rigaut, F., Graham, J. R., Jayawardhana, R., Johnstone, D., Kalas, P. G., Macintosh, B., & Racine, R. 2007, ApJ, 670, 1367-1390

Lineweaver, C. H., & Grether, D. 2003, ApJ, 598, 1350-1360

Macintosh, B. A., Graham, J. R. Palmer, D. W., Doyon, R., Dunn, J., Gavel, D. T., Larkin, J., Oppenheimer, B., Saddlemyer, L., Sivaramakrishnan, A., Wallace, J. K., Bauman, B., Erickson, D. A., Marois, C., Poyneer, L. A., & Soummer, R. 2008, SPIE, 7015, 31

Neuhäuser, R., Mugrauer, M., Seifahrt, A., Schmidt, T. O. B., & Vogt, N. 2008, A&A484, 281-291

Nielsen, E. L., Close, L. M., Biller, B. A., Masciadri, E., & Lenzen, R. 2008, ApJ, 674, 466-481

High Contrast Imaging: A New Frontier for Exoplanets Search and Characterization

R. U. Claudi, M. Bonavita, S. Desidera, and R. Gratton

INAF; Astronomical Observatory of Padova. Vicolo dell'Observatorio 5, I-35122 Padova, Italy

G. Tinetti

Dpartment Astronomy & Astrophysics, University College of London, Gower St., WC1E 68T United Kingdom

J.-L. Beuzit

Laboratoire d'Astrophysique de L'observatioire de Grenoble, BP 53, F-38041, Grenoble Cedex 9, France

M. Kasper

ESO, Karl Swartzchild Str. 2, 85748 Garching bei München, Germany

C. Mordassini

Pysikalisches Insitut, University of Bern, Sidlerstrasse 5,CH–3012 Bern Switzerland

Abstract. The discovery of 51 Peg in 1995 initiated the search for extrasolar planets with radial velocity. Since then, more than 330 expoplanets have been discovered with this successful technique. After a more timid start, the search for extrasolar planets with the transit method has begun to collect very promising results (55 planets discovered up till now). COROT, Kepler, TESS and ground-based surveys will provide many more candidates in a short term future. Moreover, for a selected sample of transiting exoplanets it is already possible to probe their atmospheres. Although very successful, both these methods are sensitive to planets which orbit quite close to their parent star. High contrast imaging will be the new frontier for exoplanet search and characterization. This technique will provide the opportunity to explore planets with masses down to the earth mass and/or orbiting at larger separation from their parent star, especially in the habitable zone. The possibility to couple an integral field spectrograph to a module for extreme adaptive optics and a 8m class telescope (SPHERE for VLT and GPI for South Gemini) or in the future to ELTs (EPICS), will allow to characterize the atmospheres of the observed exoplanets with low resolution spectroscopy. Here we present the advantages and limits of the high contrast imaging technique to detect and characterize exoplanets in the short and long term future, expecially compared to the RV and Transit methods.

1. Radial Velocity and Transits: the Present State of the Art

About ten years ago, the discovery of 51 Peg b opened the the high precision radial velocities frontier for the search of exoplanets. With this prolific technique, a number of about 330 exoplanets were discovered but, for intrinsic limits of the method, not yet characterized, except for the ones transiting their parent stars. On the other hand, after a timid start, the search for extrasolar planet transit began to harvest results (up today about 50 planets) opening the second frontier in which it's possible to perform a characterization of the atmosphere of the planet. This kind of studies are altough limited to a well defined class of objects strongly interacting with their host stars. Several focused surveys before and COROT, Kepler and PLATO and TESS in a proximate future are going to fill this sub sets of exoplanets.

Radial Velocity and transit technique are most sensitive to close in giant planets and this is due to intrinsic limits of these methods. Moreover RV searches include usually quite old stars, in order to avoid problems due to stellar activity.

2. Evaluation of the Detection Capabilities

Indirect detection techniques are going to enlarge the exo - planets candidates estate, but are not able to characterize physical properties of these bodies. In particular for intrinsic limitation of the methods no more than mass, radius and some orbital parameters can be evaluated, so every information about the atmosphere of the planets and their constituents is missing. Direct observation of photons coming from planets are strongly required in pursuing this aim. The new high contrast imaging instrumentation that are going to set up at the focus of large telescopes could open the third frontier of the planet characterization.

In order to evaluate and compare the capabilities of the high contrast imaging thecniques (e.g.: SPHERE and EPICS) with the different exoplanet search techniques (mainly RVs and transits), we made some tests using a synthetic planet population, obtained using the following assumptions:

- masses and semi-major axis are taken from the syntetic populations obtained following the Bern formation models (see Mordassini et al. 2007; Alibert et al. 2005, for details), for an host star of $1 M_\odot$;
- all other orbital elements, including eccentricity, are randomly generated assuming uniform distributions;
- intrinsic luminosity is evaluated according with the models by Baraffe et al. (2003);
- the reflected light is obtained scaling the Jupiter luminosity with planet distance and radius;
- the planet radius is computed following the approach of Fortney et al. (2007).

3. SPHERE: the Next Step

SPHERE (Spectro-Polarimetric High-contrast Exoplanet Research) is a second generation instrument for VLT devoted to exo planet finding. In a complementary way to RVs and transits research, SPHERE focuses on the outer regions of young planetary systems (Beuzit et al. 2008). As shown in Fig. 1, altought it is not able to detect the planets already discovered by the other techniques (except probably for some favourable cases e.g. Eps Eri), SPHERE will detect as many planets as the age of the system decreases, increasing the star/planet luminosity contrast [1]. Furthermore it will be able to give a first order characterization of such planets with its Integral Field Spectrograph.

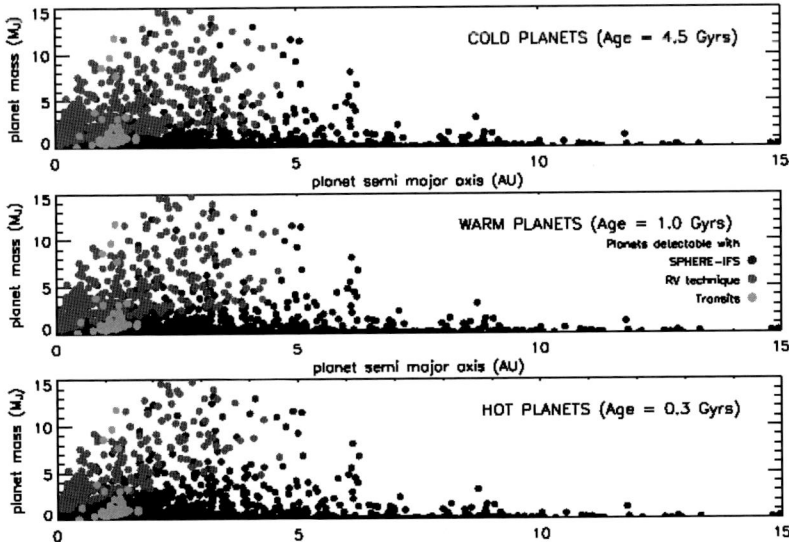

Figure 1. Comparison of the detection capabilities of different methods, assuming different values for the age of the host star. The black dots are the generated planets. The lighter shaded (coloured) dots are the planets detectable with the varius techniques.

4. EPICS: a step further

If SPHERE, and/or the other next generation instruments like GPI, will focus their attention on young planets on rather large separation from their parent stars (5-10 AU), the new E-ELT instruments will really give a chance to thake a look to a whole planetary system. EPICS (Kasper et al. 2007) will in fact be able to explore a wide range of planet masses and separations (see Fig. 2) and also to characterize the planets already discovered both with the indirect methods and with SPHERE.

---

[1] Note that the Bern's models do not take into account possible outward migration mechanisms which could feed the outern regions (a > 10-15 AU). This could significantly reduce the number of planets detectable with SPHERE-IFS.

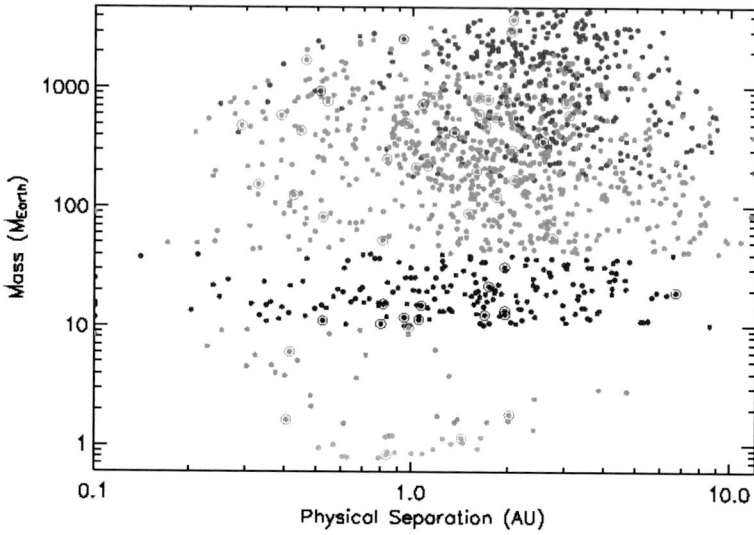

Figure 2. Semi-major axis VS mass of the synthetic planets detectable with EPICS. The dots surrounded by open circles are the planets in the Habitable Zone of their parent star.

5. Conclusions

As expected, Direct Imaging appear to be complementar to the traditional indirect methods such as Transits and Radial Velocities. In the near future, the imaging surveys will be focused on warm, giant planets, orbiting far away from theyr host stars. This will be the niche of the next generation instruments, like SPHERE and GPI. The venue of the E-ELT instruments like EPICS will give the opportunity to extend the range of detectable mass, down to some earth masses and, in some favourable cases, to see planets in the habitable zone.

References

Alibert Y., Mousis O., Mordassini C. & Benz W. 2005 ApJ, 626, 57
Baraffe, I. Chabrier, G. Barman, T.S. Allard, F. & Hauschildt, P.H. 2003, A&A, 402, 701
Beuzit, J.-L., Feldt M., Dohlen K., Mouillet D., Puget P., Wildi, F., Abe, L., Antichi, J., Baruffolo, A., Baudoz, P., Boccaletti, A., Carbillet, M., Charton, J., Claudi, R., Downing, M., Fabron, C., Feautrier, P., Fedrigo, E., Fusco, T., Gach, J.-L., Gratton, R., Henning, T., Hubin, N., Joos F., Kasper, M., Langlois, M., Lenzen, R., Moutou, C., Pavlov, A., Petit, C., Pragt, J., Rabou, P., Rigal, F., Roelfsema, R., Rousset, G., Saisse, M., Schmid, H.-M., Stadler, E., Thalmann, C., Turatto, M., Udry, S., Vakili, F., & Waters,R. 2008, SPIE, 7014, 41
Fortney, J.J., Marley, M.S., & Barnes, J.W., 2007, ApJ, 659, 1661
Kasper, M., Verinaud, C. Beuzit, J.-L. Yaitskova, N., Hubin, N., Boccaletti, A. Dohlen, K., Fusco, T. , Glindemann, A., Gratton, R. & Thatte, N. 2007, Conf. Proc. In the Spirit of Bernard Lyot: The Direct Detection of Planets and Circumstellar Disks in the 21st Century, ed. Kalas, p., (Uni. California, Berkely), 35
Mordassini, C., Alibert, Y., Benz, W., & Naef D. 2007, arXiv0710.5667

Detailed Spectroscopy of Exoplanets Using the New Worlds Observer

Webster Cash and the New Worlds Observer Team

University of Colorado, Boulder, CO 80309, USA

Abstract. It is becoming clear that spectroscopy will play an ever more central role in our understanding of exoplanetary systems. As direct imaging becomes more powerful, so must the direct spectroscopy that will follow. Here we discuss the New Worlds Observer, a mission concept for the coming decade that utilizes an external occulter (or starshade) to block out the obscuring light from the central star and leave the signal from planets free from interference and subject to spectroscopy. It is through this means that we hope to measure molecules (and biomarkers) in all kinds of exoplanets, including Earth-like.

1. Introduction

Today I stand in the Observatoire de Paris talking about how to reduce diffraction from stars to reveal faint exoplanets hidden in their glare. I am told that the blackboard behind me is the very one use by Francois Arago in his public lectures nearly two centuries ago. Arago, along with Fresnel, set the stage for understanding the wave nature of light and simultaneously provided the quantitative framework for designing new generations of telescopes. That remains true today.

The public wants to know what marvels of the Universe lie hidden over our horizons. Are there warm, watery paradises awaiting a space-faring race? Do planets everywhere harbor teeming life? Or is Earth a unique and fragile outpost of life in a vast and empty Universe? We, the science community, are poised to definitively address these questions in the coming decade.

But astronomers need more. We must follow the discoveries of what exists with experiments that determine the how and the why. How did planets throughout the universe come into being, and why are they in their current state? What are the circumstances under which life arises? Is it common or rare? And why?

To fully address these questions the science community must build facilities capable of peering into neighboring star systems with unprecedented clarity. Through direct imaging, one can find most of the major planets with just a single exposure. Through spectroscopy, one can determine the true nature of each planet discovered. Is it a barren wind-swept rock like Mars, a ferocious oven like Venus, or a watery cradle for life like the Earth?

Hundreds of planets have now been detected by ground-based radial velocity measurements. Many hundreds more (including some true Earth-like planets) will likely be detected by the Kepler Mission in the next three years (Basri et al. 2008). Space missions under design today must acknowledge that detecting the existence of Earth-like planets is no longer a goal worthy of the high price and long lead time inherent to a space observatory. Only direct spectroscopy of molecules, pushed down into the habitable zones of many dozens of planetary systems will provide the answers to the burning questions that still will remain at the forefront ten years from now.

We are completing a detailed mission study of the New Worlds Observer, as embodied in a 4m telescope flying in conjunction with a 50m external occulter that we refer to as a starshade. We have shown that the starshade technology cleanly resolves the issues of exoplanet observatory design that have arisen over the last decade. Full suppression of the starlight before it enters the telescope relieves the telescope of all special requirements such as ultra-high wave front quality correction and maintenance.

Indeed, we have shown that starshades can be used in conjunction with any telescope flying in a low acceleration environment like L2. A starshade can be designed to work with JWST with no changes to the current telescope design. The telescope need only be sufficiently powerful to resolve and study the exoplanetary system revealed once the starlight is suppressed. Having full versatility in telescope design makes the mission even more valuable, since the majority of time is spent by the starshade moving from target to target. So, most of the time, the telescope is in service to the rest (i.e., non-exoplanet part) of the astronomy community. At any given price point, starshades will provide the best science and exploration return. A starshade used in conjunction with either a small, dedicated telescope or with a pre-existing facility like JWST or JDEM can provide the detailed imaging and spectroscopy of terrestrial planets necessary to take exoplanetary astronomy to the next level. Alternative approaches to direct imaging in space like Mid IR interferometry or visible light internal coronagraphy require risky and expensive technology and yet still yield inferior data due to their inefficiencies.

The study has shown that starshades are an extendable technology. We are not on a steep technology curve where increased performance requires ever-more-difficult and subtle instruments. Where internal coronagraph systems required more than a dozen high precision optics in sequence, starshades require only a large sheet of dark plastic followed by the minimum two reflections of a conventional quality telescope. Starshades should become the exoplanetary optical accessory of choice for the foreseeable future.

Because the technology is in good shape, after some quick development efforts, a flight program can be entered with well-understood and controlled risks. The launch of a starshade could be envisioned in as little as six years. It is likely that a 4m telescope would require a few years more.

2. Concept

The famous event that led to the acceptance of the wave nature of light happened here in Paris in 1818. It was noted that Fresnel's radical new theory for the propagation of light made the prediction that if a wave of light were to pass around a circular occulter, it would cause a bright spot of light to be âĂIJfocusedâĂİ into the center of the shadow. Arago showed that the spot, which now bears his name, actually exists. But allowing light to be focused into the center of a starshade shadow is disastrous to our ends.

The idea of a starshade for exoplanetary work is not new. It goes back to at least the beginning of the space age Spitzer (1962). But the problem of light diffracting around an external occulter made the designs impractical (Marchal 1985; Copi & Starkman

2000) for revealing Earth-like planets. Three years ago, Cash (2006) found an apodization function that, for the first time, made such a system practical and affordable with today's technology.

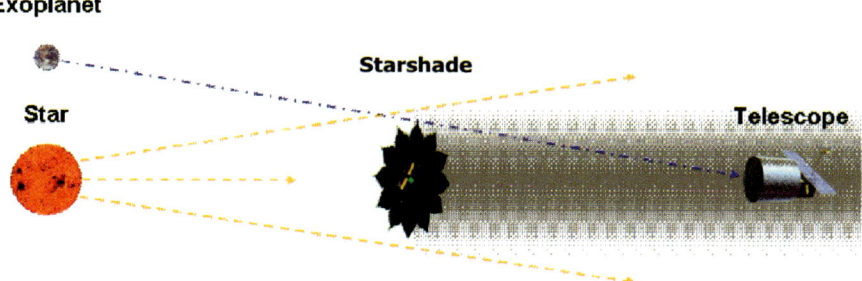

Figure 1. A starshade throws a shadow over the telescope, but the light from the exoplanet still comes over the edge of the starshade unimpeded.

An external occulter works as is shown schematically in Fig. 2 An opaque screen, larger in diameter than the aperture of telescope is flown into the line of sight from the telescope to the star. If the shade is sufficiently distant it will subtend a small angle and can blot out the star, while allowing the light from an exoplanet to slide unobscured over the edge. Geometrically, the occulter would have to be at least 5m in diameter to cast a shadow large enough to fully darken the telescope aperture. And, for a 5m object to subtend 0.2 arcsecond, it must be 5Mm (5000km) away. So the idea fundamentally requires two spacecraft flying at large separations.

But the problem of diffraction around the occulting mask, into the shadow, forces the occulter to be much larger if high contrast (like the 10^{-10} desired) is to be achieved over a small angular separation. An offset hyper-Gaussian (Fig. 2) can reduce diffraction into such a shadow by many orders of magnitude, in any given size and angular suppression range.

In order to realize this apodization function without creating undue scatter from transmitting materials, the starshade must be made binary âĂŞ that is, the starshade

$$A(\rho) = 0$$
For $\rho < a$ and

$$A(\rho) = 1 - e^{-\left(\frac{\rho - a}{b}\right)^n}$$
For $\rho > a$

Figure 2. The apodization function that allows the starshade to fully suppress the diffraction.

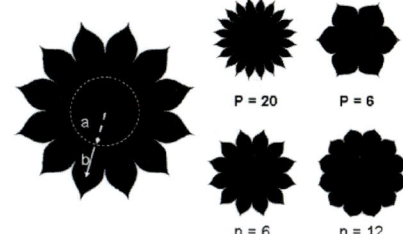

Figure 3. The shape of the starshade changes as a function of the parameters

must be everywhere fully opaque or fully transmitting. In practice this means the starshade will resemble a flower with petals as shown in Fig. 3.

During the course of this study we showed that the minimum number of petals to maintain the diameter of the central, deep shadow is sixteen. However, one can use as few as twelve petals with only a modest loss of shadow diameter.

In the course of this study we settled upon a set of baseline mission architecture parameters, given in Table 1. This starshade design represents a balance between size and cost on one side and Inner Working Angle (IWA), and long wavelength limit on the other.

The shade is 50m in diameter to the petal inflection points and 62m tip-to-tip. It is made of opaque plastic and is not an optic in the conventional sense of the word. It is only the projected outline onto the sky that determines its performance.

We have developed three physical optics simulation codes that allow us to predict the shadows quantitatively. Fig. 4 shows the suppression efficiency of the baseline starshade design as a function of both radius and angular offset for two nominal wavelengths at the nominal separation of 80,000 km. It is clear that, throughout the visible band, the central star is fully suppressed while the exoplanets as close in as 40 mas are fully visible.

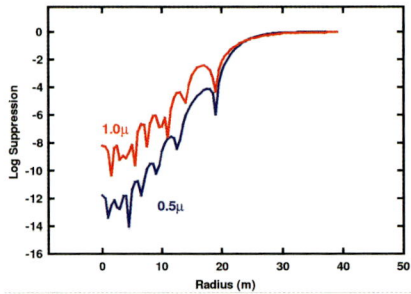

Figure 4. The diffraction pattern falls by over ten orders of magnitude across just 25m. This supports observations of planets as close as 40mas from the central star.

Table 1

Baseline Parameters	
a	12.5m
b	12.5m
n	6
P	16
R_{max}	31m
F	80,000km

We have shown in our study that the starshades can be used in conjunction with any conventional telescope in a low acceleration orbital environment like the Sun-Earth L2 point. But the size of the telescope makes a major difference in what can be observed. Fig. 2 shows a series of simulations of our Solar System viewed pole-on from a distance of 10pc with a starshade blotting out the central star. As the diameter of the telescope increases from left to right, the exoplanets emerge from the confusion. The diffraction limit on a telescope determines its resolution and hence the quality of observation on a distant system. A 10m telescope, returns truly spectacular images with Earth-like planets leaping off the page. However, with a telescope of only one meter aperture, the resolution has fallen to the point where individual planets cannot be resolved from each other, and many planets would be difficult to separate from the fog of exozodiacal light. We chose to study a 4m telescope as that is the largest that can be readily built in the coming decade.

Exoplanet Spectroscopy with the New Worlds Observer 213

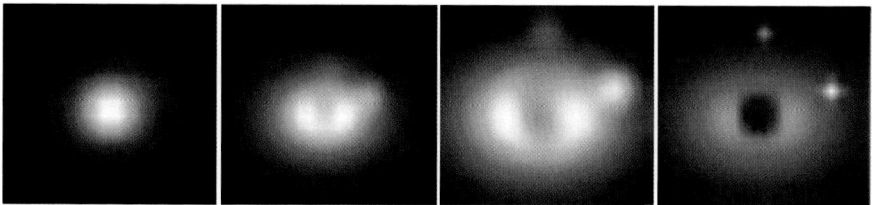

Figure 5. The 4m diameter of the telescope is not set for reasons of inner working angle or collecting area. It is needed to cleanly resolve the features of a planetary system once it is revealed.

Detailed engineering of the starshade systems and the telescopes to be used with them is now underway and is beyond the scope of this talk. In Fig. 6 we show a CAD drawing of a starshade with its spacecraft. One can see the very large solar panels needed to run the ion electric propulsion engines that move it from target to target.

3. Science

The science enabled by the New Worlds Observer is huge, even for a Flagship Class space-borne observatory. We have established that about 70% of the mission timeline will be spent with the starshade moving from target to target and with the telescope free for the full range of astrophysical research. The telescope will be a true successor to HST, covering the same band of the spectrum from Lyman-α to the near IR. The resolution will improve by a factor of four over much of the visible and the ultraviolet. It would have over twice the collecting area, higher observing efficiency, wider field of view, and better detectors yielding an order of magnitude more data.

The New Worlds Observer will find exoplanets âĂŞ lots of exoplanets. If NWO were to observe our system from 10pc it would immediately detect seven planets (all but Mercury) including three in the habitable zone. NWO will be able to study about twenty-five systems per year for five years. So if our system proves to be average, then NWO could discover nearly a thousand exoplanets, over three hundred of which would be in the habitable zone.

Much has recently been made of the need for an astrometric mission precursor to the New Worlds Observer. The idea, as has been stated, is that a starshade is clumsy, moving slowly from system to system and therefore knowing which systems to observe and when would greatly enhance the efficiency and productivity of the mission. During this study we have determined that NWO is also the best way to find planets, particularly Earth-like ones. Its efficiency and sensitivity are so high that it more than makes up for the slow pointing. A system need only be visited once or twice to detect all major planets and to determine immediately the nature of the planets. While knowing the mass of a discovered planet is nice, it is not as important as a spectrum. Furthermore, the astrometric missions would give masses for only a small fraction of the many detected planets.

In this study we looked at all of the nearby stars and analyzed the likelihood of finding Earth-like planets. The full analysis will shortly be submitted for publication (Turnbull et al. 2009). Fig. 7 is a plot of the contrast of Earth relative to parent star as a function of the angular size of the habitable zone (HZ). The vertical lines show the IWA

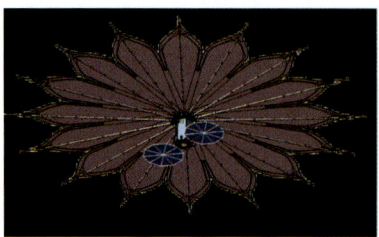

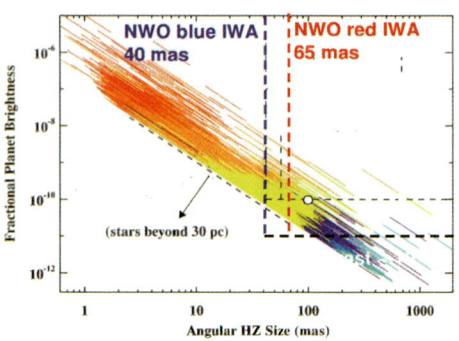

Figure 6. CAD drawing of the starshade with its spacecraft.

Figure 7. All the habitable zones of the nearest stars are shown as a function of angular diameter and the amount of suppression needed to see an Earth. NWO will mostly detect Earths around early K to F stars.

of NWO in blue and in the red. One can see that most of the targets are early K-type or earlier. By the time one reaches A stars, the relative brightness of an Earth is much lower and NWO might experience difficulties. But, an HZ planet around an OB or A star might not be considered habitable anyway.

We took all the star systems that could be observed in the habitable zone and calculated the probability of seeing a planet therein during the first visit. We then ordered them by probability and plotted them in the histogram of Figure 8. If there is, on average, one planet per HZ, then each bar represents the expected number of HZ planets for that system.

The sum of the bars equals the total number expected to be found in the mission. It is important to note that by the time one has reached the 100th best target system, the probability has fallen to under 10 percent. Those systems will be so small angularly that only planets in the outer regions of the HZ that happen to be at quadrature will be seen. In short, one is scraping the bottom of the barrel. Since NWO can observe over 100 lines of sight before refueling the starshade, NWO runs out of good targets before it runs out of fuel. If Earth-like planets prove to be common, then there will many dozens to study in detail from those best star systems.

Well before launch results from the Kepler mission will tell us the needed statistic. If (as seems unlikely right now) Earth-mass planets in the HZ are determined to be rare, then there is an adjustment to be made to the mission design. Figure 8 changes dramatically with the IWA. For example, if the IWA is reduced to 40m as from the 62m as above, the expected number of HZ planets triples. This can be accomplished by working only in the blue with a smaller shade, or by working with a larger shade that takes longer to travel between systems but provides excellent probability of detection upon arrival. The NWO mission concept is the only one currently under study that can adjust to the reality of the Universe in this way.

It is, of course, once the planet is seen, that NWO really begins to shine. NWO will carry an Integrated Field Spectrometer with a sufficiently wide field that it can provide spectra of the entire HZ at a single shot. Special channels will be adjustable so that multiple outer planets can be studied at the same time. With one to three days on target, high-quality spectra of all the major planets will be captured. A simulated spectrum of the Earth at 10pc, viewed for about one day by NWO, is shown in Figure 9. All known sources of noise are included. It is a truly exciting prospect that such data could be available in under ten years! Clearly visible in the spectrum is the rise to short wavelength, indicating a blue planet. Toward the red end are strong absorption features of water, indicative of oceans and clouds. Even more exciting is the presence of absorption lines from molecular oxygen and an absorption edge from ozone in the near ultraviolet. These features are in the spectrum of the Earth solely as a byproduct of plant life.

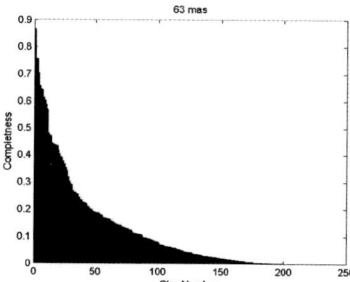

Figure 8. The probability of seeing a planet in the habitable zone is shown for all the best candidate systems for an IWA of 63mas.

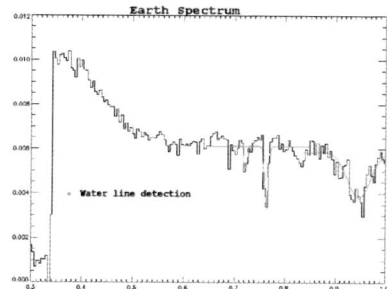

Figure 9. The spectrum of the Earth as seen by the NWO. Note the prominent water and oxygen absorption lines and the ozone edge in the near UV.

The study of exoplanets through imaging and spectroscopy is just the beginning. We can learn a great deal more through looking at the systems as a whole. In fact, we must give a great deal of priority to the diffuse emission from the planetary system. The exozodiacal light and debris disks will be much clearer in NWO images than in any other mission concept under study. Indirect methods are entirely insensitive to disk structures. Internal coronagraphs leave residual speckles and scattered light that must be subtracted out, largely hiding extended structures. But observing exozodis is crucial, both as a help and as a hindrance. Luckily, NWO is robust against the presence of bright exozodis. Yet we know exactly nothing about zodiacal light in other systems. It could be that our home system, having an asteroid belt, has unusually high levels of interplanetary dust, and that our view of many neighboring systems will be unimpeded. Or it is equally possible that we have a naturally cleaner system than most. Measuring this effect is crucial to understanding the future of direct observation of exoplanets.

Yet exozodiacal light also provides a treasure trove of scientific discovery. The distribution of the light is a sensitive tracer of the systemâĂŹs orbital dynamics. Planetary orbital resonances will be displayed as gaps and enhancements in the dust. Tiny planets, too small to be seen directly will leave distinct marks. The dust seen gives us critical information like the inclination and orientation of the systemâĂŹs ecliptic plane. Figure 9 is a simulation showing a hypothetical system with three planets âĂŞ Venus, Earth and Jupiter. The exozodiacal light has total brightness equal to our own, but has been made more extended for effect.

By eye, one can place an ellipse over the system, estimating accurately the orientation of the plane. Then concentric ellipses may be drawn about the central star. Those that pass through a planet show the orbit under the assumption of circularity. Exozodiacal light has the potential to give us the orbit of each planet from a single image!

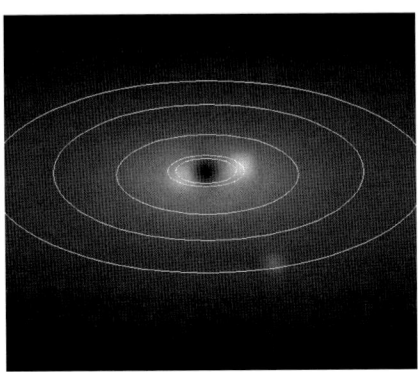

Figure 10. The exozodiacal light can ppinpoint the inclination of the system and therefore the orbits can be immediately determined.

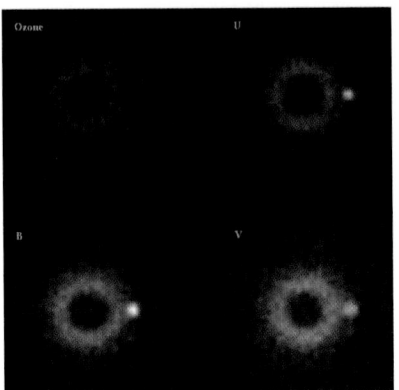

Figure 11. Simulation of the solar system seen pole-on in the four bands of the ExoCam simultaneously. Note that it is obvious that the Earth is blue and that there is an ozone edge.

The full suite of astrophysical techniques will be available to the observer. Polarimetry will play a prominent role in analysis of the systems. The light we see from planetary systems in the visible band is generated in the (now obscured) star and scatters off the planets and debris. Thus we expect very high levels of polarization (Stam et al. 2004, 2008). Similarly, other GA instruments might be used in special circumstances. For example, a high resolution spectrometer might be used to capture a spectrum of a particularly interesting planet.

We studied a variety of focal plane instruments in order to evaluate the likely system performance. The first instrument is a camera, which we refer to as ExoCam. It consists of a dichroic string that splits the 250-1700nm light into four bandpasses for immediate color imaging (Pale Blue Dot!) and broadband photometry. A simulation of the Earth as observed with ExoCam is shown in Figure 13. The detectors will be photon-counting CCDâĂŹs which provide the best signal to noise.

The New Worlds Observer is the right tool for finding exoplanets and establishing their nature. It shows a technical path to full and detailed study of molecules on exoplanets.

Acknowledgments. I am summarizing here the work of many individuals who have participated in the New Worlds Observer study. I wish to thank them all for helping fill in the details of how to make a starshade work and how to use it. In particular I wish to thank M. Turnbull, T. Glassman, A. Lo, and P. Oakley for use of their results in advance of publication. This work was supported by NASA Grant NNX08AL70G.

References

Basri, G., Ramos-Stierle, F., Soto, K., Lewis, T., Reiners, A., Borucki, W., Koch, D. 2008, "The Kepler Mission: Terrestrial Extrasolar Planets and Stellar Activity", 14th Cambridge Workshop on Cool Stars, Stellar Systems, and the Sun ASP Conference Series, Vol. 384, Ed. Gerard van Belle., 281.
Cash, W., 2006, Nat, 442, 51-53
Copi, C. and Starkman, G., 2000, ApJ, 532, 581.
Marchal, C., 1985, Acta Astronautica, 12, 195
Oakley, P., Cash, W., & Turnbull, M., 2008, Proc. of SPIE, 7010
Oakley, P. & Cash, W., 2009, ApJ, submitted
Spitzer, L., 1962, American Scientist, 50, 473
Stam et al., 2004, A&A, 428, 633-672
Stam et al., 2008, A&A, 482, 989-1007
Tinetti, G., et al., Astrobiology, 3(4), 2005
Tinetti, G., et al., Astrobiology, 6(6), 2006
Turnbull, M., et al., 2006, ApJ, 644, 551
Turnbull, M., et al., in preparation, 2009.

Spectral Analysis of Atmospheres by Nulling Interferometry

M. Ollivier and S. Jacquinod

Institut d'Astrophysique Spatiale - Université de Paris-Sud 11 and CNRS (UMR 8617)

Abstract. Nulling interferometry has often been considered as one of the most promising techniques to get spectra, in the thermal infrared spectral range, of exoplanet and particularly telluric ones. The required performance, in terms of nulling depth, spectral bandwidth and stability begins to be achievable in the laboratory. However, concrete observatory projects have strong difficulties to emerge in space agencies programs, because of mission requirements complexity. In this paper, I present the performance various scientific objectives need and the state of the art of this technique, including mission aspects. I finally propose elements that could help defining a first space mission project.

1. Nulling Interferometry : Theoretical Aspects

1.1. Principle of Nulling Interferometry

Nulling interferometry has been proposed in the late 70s as a possible technique to detect and spectroscopically characterize planets around nearby stars at wavelengths where diffraction prevents from the use of single aperture telescopes (Bracewell 1978). The principle of nulling interferometry in a 2-telescope configuration is described in figure 1 (left). The idea is to use two small telescopes, pointed in the direction of the star and to combine their beams. On the beam combiner the wavefronts from both star and planet create interferences. If we introduce a π phase-shift in one arm of the interferometer, the stellar interference is destructive (the optical path difference is equal to zero). In the direction of the planet, and because of the angular distance to the pointing direction, the path difference is not equal to zero. Adjusting the distance D between the two telescopes, one can obtain that the path difference ($D \times sin(\theta)$) introduces a phase-shift that compensates the instrumental one at the observation wavelength. In that way, one gets a constructive interference in the direction of the planet. Thus, the distinction between the stellar and planetary photons is performed by their phase difference. The transmission map of such an instrument is shown in figure 1 (right).

1.2. Key Issues of Nulling Interferometry

The main characteristics of a nulling interferometer are :

- The depth of the null, i.e. the capability of the instrument to cancel the stellar light. This parameter is quantified by the rejection rate, which is the ratio between off-axis and on-axis transmission of the instrument (ratio between constructive and destructive interference). The null depth is also characterize by the reverse of the rejection rate.

- The chromatism of the null, i.e. the spectral range over which the rejection rate is performed,

- The stability of the null, i.e. the rms amplitude of rejection rate fluctuation, as a fraction of the rejection rate.

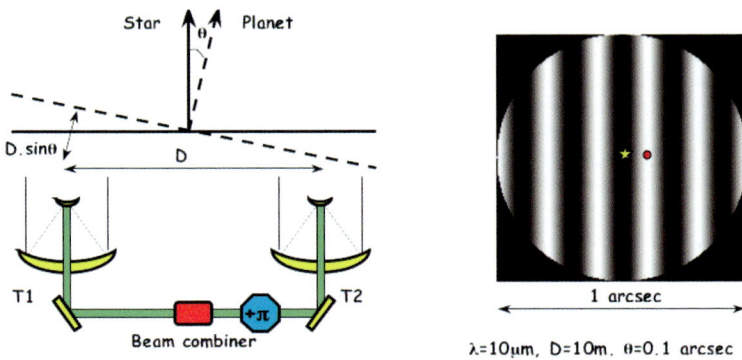

Figure 1. (left) : principle of nulling interferometer. See the text for description, (right) : transmission map of a two-telescope nulling interferometer

The depth of the null directly leads to the level of stellar background that will be seen during the stellar companion observation, and as a consequence to the limiting contrast of the astronomical scene. The chromatism of the null conditions the spectral range and the spectral width of the observation, and as a consequence, the amount of photons that can be used for detection or spectral analysis. The stability of the null conditions directly the noise level of the signal and the ability or not to disentangle the stellar leaks from the companion signal itself.

Getting an optimal null requires a perfect matching of the wavefronts to be recombined. For instance, in the case of a 2-telescope interferometer, requirements can be given in terms of :

- amplitude : If I_0 is the intensity of each beam on the beam combiner and δI is their intensity mismatch, then the rejection rate ρ can be computed as,

$$\rho = \frac{I_{max}}{I_{min}} = \frac{I_0(2 + \delta I) - 2I_0 \sqrt{1 + \delta I}}{4I_0} \simeq \frac{\delta I^2}{16} \quad (1)$$

- phase : if $\delta\phi$ is the phase mismatch between the wavefronts on the beam combiner, then the rejection rate can be computed as

$$\rho = \frac{I_{max}}{I_{min}} = \frac{2I_0 + 2I_0 cos(\pi + \delta\phi)}{4I_0} \simeq \frac{\delta\phi^2}{4} \quad (2)$$

- polarization : polarization effects are more complex to described. They have be studied by many authors in many contexts (see for instance Chazelas et al. 2006).

In order to monitor and control the amplitude, phase, and polarization effects, a nulling interferometer is usually made of several optical devices :

- optics for beams transportation,
- a fine relative angle sensor and associated correcting actuators,
- optical delay lines to equalize the optical path difference between the beams at the level of a few nanometers,
- an achromatic phase shifter device to cancel the stellar light at every wavelength,
- a differential beam intensity matcher,
- a differential beam polarization matcher,
- optical filtering stages,
- a beam combination stage,
- a detection stage.

1.3. From Optical Principle to Observatory Concepts

The intitial concept of Bracewell has been developed particularly in the thermal infrared spectral range, to take into account the specificities of telluric planets detection :

- the size of the planet : the radius distribution of telluric exoplanet has to be determined because the planet size (and thus the signal level) is a strong constraint on the size of collecting apertures;
- the angular distance from the planet to the star, varying with time (position of the planet on its orbit, relative inclination of the orbit plane with respect to the observer ...);
- low level signal from the planet, typically several tens of photons per second and square metre in the [6-20 μm] spectral range for an earthlike planet at 10 pc;
- presence of zodiacal emission in the solar system : in the thermal infrared, the contribution of our zodiacal cloud is several hundred times higher than the thermal emission of the Earth;
- presence of a zodiacal cloud around the exo-system : at present the level of exo-zodiacal level is not known for nearby stars, but has to be estimated because it is one of the dimensioning parameters for a characterization mission either in the visible or in the thermal infrared spectral range;
- instrumental noise associated to the instrument temperature : in the thermal infrared, the level of the integrated local zodiacal emission and the thermal emission of the instrument reachs several hundred to thousand times the level of the signal to be detected;

- the level of stellar leaks : taking into account the way beams are combined, the mean stellar leaks can reach 10 to 100 times the signal level. A trade-off between this level of stellar leaks and the beam combiner complexity has to be found.

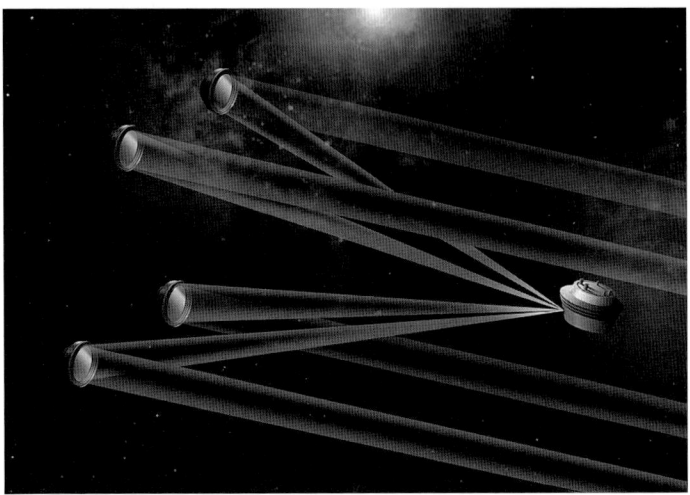

Figure 2. Principle of the EMMA concept of nulling interferometer. The telescopes are located on the parabola and the beam combiner is put at the focus (source : NASA-JPL)

All these constraints lead to concepts with several telescopes :

- free floating in space (formation flying) in order to allow a fast configuration of the array with respect to target distances and sources characteristics (distance to the star, position on the orbit ...); rigid (deployable) structures with telescopes on fixed length beams are not considered any longer because of their lack of versatility;

- increasing the size of the nulled transmission area on the sky (Angel et al. 1986; Rouan 2003) and the mean depth of the null : this point is particularly important to reduce the level of stellar leaks, taking into accound the non-zero size of the stellar disk.

- allowing the creation of sub arrays and partial beam combination between sub-arrays (Mennesson et al. 2005),

- internal modulation between sub-arrays, allowing fast planetary signal modulation and lock-in detection at a frequency much higher than frequency allowed by only rotation of the whole array (Mennesson et al. 2005; Absil et al. 2003). This point allows disentangling the planetary signal (non centro-symmetric signal with respect to the pointing direction) and the exo-zodiacal background (centro symmetric with respect to the pointing direction). Present mission concepts are nammed TPF-I at NASA (Coulter 2003; Lawson et al. 2007) and DARWIN at

ESA (Karlsson & Kaltenegger 2003). The are both based on the EMMA concept (telescopes located on a parabola, with the beam combiner at the focus see figure 2. This concept is a good trade-off between many constraints and requirements (optical and thermal aspects, baseline versatility, imaging capabilities ...)

Whatever the configuration of the array and the beam combination method can be, the signal reaching the detector array is the sum of many contributions : the signal from the planet(s) (modulated), the mean stellar leaks (not mudulated), the mean exo zodiacal light contribution (not modulated because of clever interferometer configuration), the solar system mean zodiacal light contribution (not mudulated), the fluctuation of stellar leaks linked to the instrument stability and asymmetry of the astronomical source (variable contribution, that can be considered as noise), the variable contribution of the exo zodiacal light (linked to the cloud asymmetry, modulated by the internal modulation), and other sources of noise (thermal, instrumental and detection noises) that affect all the components mentioned previously. Huge data processing is thus required to disentangle all the contaminants from the planetary signal itself. Complex algorithms that take into account the spectral nature of the signal and multi baseline information have been proposed in that purpose (Mugnier et al. 2006). They aim at providing both maps of the planetary systems and a spectrum of each planet over the observational spectral range, with a spectral resolution of several tens, depending on the magnitude of the target.

2. Nulling Interferometry in the Laboratory

Nulling interferometry is theroretically a very interesting technique. Practically, the requirements on wavefront quality, in terms of differential amplitude, phase and polarisation on each point of the beam combiner lead to strong requirements on optical components and sub-systems themselves. In order to reduce optical pieces and coatings manufacturing requirements, several concepts have been proposed such as optical filtering by a pinhole (Ollivier & Mariotti 1997) or by single mode waveguides (Mennesson et al. 2002) to clean the wavefront from medium to high order defects, symmetric beam combination (Serabyn & Colavita 2001) to allow a perfect treatment of each polarisation, interferometric achromatic components such as achromatic phase shifters to allow observation in a broad spectral range and spectral analysis (Serabyn et al. 1999; Baudoz et al. 1998), and other devices.

In order to validate the principle of such instruments, several R&D programs, and associated test benches have been successively developed in the United States and in Europe with several goals :

- to reach deep null in a narrow band in the thermal infrared and validate the concept of optical filtering by a pinhole (Ollivier et al. 2001);

- to reach deep null in a broad band in the visible spectral range using an achromatic phase shifter (Wallace et al. 2000);

- to reach deep null in a narrow band in the near infrared using a multiaxial beam combiner based on integrated optics (Barillot et al. 2004);

- to reach deep null in a narrow band in the near infrared using a symetric beam combiner (Ergenzinger et al. 2004);

- to test optical components designed specifically for nulling interferometry, such as achromatic phase shifters (Gabor et al. 2008a; Gappinger et al. 2009);

- to reach deep null in a broad band in the near infrared and to stabilize it over several hours (Gabor et al. 2008b);

- to perform an end to end simulation of nulling interferometry detection, signal processing and planetary detection (Booth et al. 2008).

At present, the best performance in terms of null depth and spectral band width has been obtained in the [8-12 μm] spectral range (Bandwidth = 50%, rejection rate = 5×10^4) with an adaptive nuller (Peters et al. 2008). As a comparison DARWIN/TPF-I type projects require a rejection rate of about 2×10^5 and a spectral bandwidth of 60 %. As mentioned above, the stabilization of the null is also a key-point of the technique to allow long time exposure. In order to spectroscopically characterize an earthlike planet at 1 A.U. from a sunlike star at 10 pc, the root mean square of the nulled signal should not be higher than 3×10^{-9} over 10 days, at an observation wavelength of 7 μm, i.e. about 10^{-4} time the mean level of the null (Gabor et al. 2008a).

3. Towards a Nulling Interferometer in Space

The first statistical observations of known exoplanets, based on 340 objects (15 Feb 2009), show a strong diversity among them (Schneider 2009) in terms of planet characteristics (size, mass, orbital semi major axis and eccentricity ...) and stellar parameters (stellar type, luminosity class, age, metallicity, activity ...). A space mission dedicated to the spectroscopic analysis of extrasolar planets should thus be versatile, or, at least, be optimized to allow the observation of several classes of targets (e.g. in given range of angular distance to the star or in a size range). Table 1 shows the observational parameters of several classes of extrasolar planets orbiting around a solarlike star.

Table 1 shows clearly that it is not reasonable to design a concept that can be optimized for all classes of objects. Taking into account the specificity of nulling interfermetry (high angular resolution and high contrast observation capabilities), one can see that the method is particularly suited for 2 classes of objects, with 2 types of instrument performance (see Table 2) :

- Hot Jupiters : in this case, required nulling performance is limited and the observatory takes benefit from high resolution capabilities of interferometry to disentangle the stellar and planetary signal. The optimal spectral range in terms of contrast (and thus instrumental complexity) and thermal environment is the L band (around $\lambda=3.5\mu$m). This concept is particularly suited to study the stellar environment of potential targets in order to determine the accurate level of their exozodiacal emission. This concept is later called "Preparatory Science Concept";

- Super and Real Earths : in this case, the required performance is stronger than in the previous case. The best spectral range is where the contrast is minimum and atmospheric spectral features are detectable, i.e the thermal infrared (around 10 μm). This concept is later called "Characterization of Earths Concept"

Table 1. Observational characteristics of several exoplanets classes orbiting around a solarlike star. A "Hot Jupiter" is a giant planet in the close vicinity of its parent star (less than 0.1 A.U) and can also be called an "Irradiated Jupiter" (and not necessary a young object), a "Super Earth" is supposed to be an object of 6 to 10 Earth masses (1.5 to 3 Earth radii) at 1 A.U., a "Real Earth" is the Earth at 1 A.U, a "Real Jupiter" is Jupiter at 5.2 A.U, and a "Uranus" is a Uranus class object at the distance of Uranus (about 20 A.U.)

		Hot Jupiter	Super Earth	Real Earth	Real Jupiter	Uranus
Distance to the star (mas)		5	100	100	520	2000
Telescope diameter (m)	0.55 μm	20	1	1	0.2	0.05
	3.5 μm	150	7	7	1.4	0.4
	10.2 μm	400	20	20	4	1
Planetary flux (mJy)	0.55 μm	0.86	3×10^{-4}	3×10^{-5}	5×10^{-5}	5×10^{-7}
	3.5 μm	2.16	10^{-4}	10^{-5}	1.5×10^{-5}	1.5×10^{-7}
	10.2 μm	1.1	1.3×10^{-3}	10^{-4}	3×10^{-5}	2×10^{-8}
Contrast star / planet	0.55 μm	35000	10^{8}	10^{9}	6×10^{8}	7×10^{10}
	3.5 μm	4000	10^{8}	10^{9}	6×10^{8}	7×10^{10}
	10.2 μm	1200	10^{6}	10^{7}	4×10^{7}	7×10^{10}

Table 2 shows clearly that a Preparatory Science and an Earthlike planet spectroscopic missions exhibit two types of complexity. The first one is what can be considered as the simpliest nulling interferometry space mission - a precursor mission - with moderate performance and scientific capabilities.
At present, only the first concept has been studied at a 0-level using space agencies standards :

- at NASA, in the framework of the FKSI mission concept study : FKSI is Bracewell nulling interfermeter with fixed baseline (12 m), optimized for exo zodiacal clouds around nearby stars characterization (Danchi et al. 2008)

- at CNES (French space agency) in the framework of the PEGASE mission concept study : PEGASE and FKSI are similar in terms of scientific goals, instrumental concept, characteristics and spectral range, except that PEGASE is based on free flyers with separate siderostats and beam combination spacecraft, each capable of its own positionning and pointing with respect to the other parts of the array (Ollivier et al. 2009). The interferometric array would be in orbit around the Lagrangian point L2.

Technical readiness level (TRL) have been estimated at both instrumental and mission level for a preparatory science mission (see Tables 3 and 4). These tables clearly show the difficulty to design, at present, a nulling interfermetry mission based on free flyers. It is not only the technology at the level of the instrument that is limited but also at the satellite and mission levels. R&D programs focused on the instrument development in the laboratory is necessary but not sufficient because contrary to other

Table 2. Main characteristics of mission concepts in the case of a "Preparatory Science Concept" mission and an "Characterization of Earths Concept" exoplanet spectroscopy mission

Characteristics	Preparatory Science	Characterization of Earths
Number of Telescopes	2	4
Spectral Band	3-7 μm	7-20 μm
Co-phasing Accuracy (residual opt path diff)	2.5 nm rms	3 nm rms
Mean rejection rate	10^4	10^5 to 10^6
Nulling stability around mean null	10 %	10^{-4} at 7μm over 10 days
Co-phasing accuracy	5×10^{-3} rad	10^{-3} rad
Baseline	10-500 m	20-500 m
Satellite guiding Accuracy	a few arcsec	a few arcsec
Fine guiding	20 mas	8 mas
Telescope size	30-40 cm	1-3 m
Instrument temperature	100 K	40 K
Detector temperature	55 K ± 1K	10 K ± 0.1 K

Table 3. Technological Readiness Level (TRL) for state of the art technology on mission aspects of a PEGASE like mission

Equipement	TRL
Thrusters	4
Radio frequency metrology	4
Optical sensor Free-flying, Guiding, and Navigation Control	4
Accurate star tracker	4
V-grooves techiques	8
Operation at L2	7

mission concepts, payload and spacecraft cannot be considered separately because of their permanent interaction. For instance, the fine metrology level required to co-phase the array is performed by the instrument itself (optical path difference measurement and correction by a fringe tracker and delay lines). As a consequence, the global TRL of a mission is limited either by the instrument TRL or the mission TRL or both.

At present, no formation flying concepts has ever been launched. It is thus very difficult to estimate correctly either the difficulties and performance or the cost of a complex array such as what is proposed for DARWIN / TPF-I. A tentative estimation of the cost of a "low cost nulling interferometer" have been done in the case of PEGASE

Table 4. Technological Readiness Level (TRL) for state of the art technology for payload aspects of a PEGASE like mission

Equipement	TRL
Infrared detectors (HgCdTe array)	9
InAsGa Photodiodes	9
CCD or High accuracy sensors	9 at usual temperature 4 at 90 K
Piezo systems	9 at usual temperatures 4 at 90 K
Delay lines	5
Achromatic phase shifters	4
Cesic benches	7
Symmetric beam combiners	3
Optics	9
Single mode fibres (optical filtering)	4

study at CNES leading to a projects of several hundred millions euros considered by ESA as very costly compared to the potential scientific return at last call for proposals for the Cosmic Vision program.

4. Conclusion : The Paradigm of Infrared Spectroscopic Observatories or "The end justifies the means philosophy"

Spectroscopic observation of earthlike planet both in the thermal infrared and visible spectral ranges has been considered as the most powerful strategy to get information on atmospheric composition of these planets and on the potential presence of life by detection of biomarkers (Selsis et al. 2008). Spectroscopic observations in the thermal infrared should be performed from space, because of atmospheric absorption. Taking into account the constrast and the angular disntace between star and planet, only free flyer interfermeter concepts appear to be adapted and versatile enough to allow the observation of several candidates in several geometrical configurations. However, the technology states of the art allowing such concepts is not sufficient and leads, at least in a first time, to costly concepts with respect to their scientific potential return. Such concepts are however necessary to get the capability to design much complex missions and particularly spectroscopic missions for earthlike planets.

To face this difficulty we suggest a "the end justifies the means" philosophy which can have several aspects :

- substantial efforts could be done to allow the development of complex observatories, whatever their real first astronomical potentials are (concept of technological missions),

- substancial efforts could be done to allow ground based developments and testing, either for preparatory science and technology. Such efforts can lead to concepts such as ALADDIN, a nulling interferometer concept to study stellar environments from the Antartica (Coudé du Foresto et al. 2007).

In any case, the present funding levels and structures are not well adapted to such big projects. A dedicated structure is maybe necessary to allow the development of instruments that may give the first elements of answer to one of the oldest question raised to humanity : Are we alone in the Universe?

References

Absil, O., Karlsson, A., & Kaltenegger, L. 2003, SPIE 4852, 431
Angel, J.R.P., Cheng, A.Y., & Woolf, N.J. 1986, Nat, 322, 341
Barillot, M. et al., 2004, Proc. ICSO 2004, ESA-SP 554, 231
Baudoz, P. et al., 1998, SPIE 3353, 455
Booth, A.J., Martin, S.R., Loya, F., 2008, SPIE 7013, 11
Bracewell, R.N. 1978, Nat, 278, 780
Chazelas, B. et al., 2006, Proc. IAU Colloquium #200, 251
Coudé du Foresto, V. et al., 2007, Proc. In the Spirit of Bernard Lyot: The Direct Detection of Planets and Circumstellar Disks in the 21st Century. ed. Paul Kalas, (University of California, Berkeley, CA, USA).
Coulter, D.R. 2003, ESA-SP 539, 47
Danchi, W.C. et al., 2008, SPIE 7013, 16
Ergenzinger, K., et al., 2004, Proc. ICSO 2004, ESA-SP 554, 223
Gabor, P. et al., 2008a, A&A, 483, 365
Gabor et al., 2008b, SPIE 7013, 70134O-70134O-11
Gappinger, R.O. et al., 2009, Appl.Optics, 48, 868
Karlsson A. & Kaltenegger, L. 2003, ESA-SP 539, 41
Lawson, P.R., et al., 2007, JPL Publication 07-1
Mennesson, B., Ollivier, M., & Ruilier, C. 2002, JOSA-A, 19, 596
Mennesson, B. Léger, A. & Ollivier, M. 2005, Icarus, 178, 570
Mugnier, L., Thiébaut, E., & Belu, A., 2006, EAS Pub Ser, 22, 69
Ollivier, M., Mariotti, J-M. 1997, Appl.Optics, 36, 5340
Ollivier, M. et al., 2001, A&A, 370, 1128
Ollivier, M. et al., 2009, Exp. Astron.,
    http://www.springer.com/astronomy/journal/10686
Peters, R.D., Lay, O.P., Hirai, A., & Jeganathan, M. 2008, SPIE 7013, 11
Rouan, D. 2000, ESA SP-539, 565
Schneider, J. 2009, http://exoplanet.eu/
Selsis, F., Paillet, J. & Allard, F. 2008, in Extrasolar Planets, (Cambridge University Press, Cambridge, UK), 245
Serabyn et al., 1999, Appl.Optics, 38, 7128
Serabyn, E. & Colavita M.M. 2001, Appl.Optics, 40, 1668
Wallace, K., Hardy, G., & Serabyn, E. 2000, Nat, 406, 700

A Spectroscopic Method for Direct Detection of Exoplanets.

Patricio Cubillos and Patricio Rojo

Departamento de Astronomía, Universidad de Chile, Casilla 36-D, Santiago, Chile

Jonathan Fortney

Department of Astrophysics, 211 Iterdisciplinary Sciences Building, UCO/Lick Observatory. University of California, Santa Cruz, CA 95064, USA

Abstract. In this work we have revisited a method to directly search the signature of a known non-transiting extrasolar planet in the IR spectrum of a star-planet system. We present preliminary results on the planetary system HD 217107 observed with Phoenix spectrograph in 2007, and also present an optimized strategy to maximize our future success considering the best conditions and best candidates to observe after performing simulations to predict sensitivity limits of the method. With positive results this method could add new information of nontransiting exoplanets and validate the high resolution models of their atmospheres.

1. Introduction

Over the last years there have been an incredibly fruitfulness in the search of exoplanets. Despite having over 320 known planets to this date through the use of several different methods, atmospheric characterization has only been possible for a handful of them. Most of them are transiting planets, which only account for around 15 % of the total population.

A successful result of the method will yield the relative flux and the inclination of the system. Together with the radial velocity measurements, these will allow us to precisely constrain the mass of the planet.

2. Observations and Data Analysis

We have observed a total of 15 hours of the system HD 217107 between August and November of 2007 using the Phoenix high resolution spectrometer at Gemini South Observatory. A total over 950 frames were obtained in 11 sets of continuous observations through a night at varied observing conditions. The spectral range covered was from 2.136 to 2.145 μm. We extracted the spectra from the frames using an IDL implementation of the Optimal Spectrum Extraction algorithm described in Horne (1986)[1]. We

---

[1] http://physics.ucf.edu/~jh/ast/software/optspecextr-0.3.1/doc/index.html

then masked out those wavelengths channels that varied the most as they are attributed to telluric absorption.

Inspired by the work of Wiedemann et al. (2001), we search the planetary signature in the spectrum using a correlation function between the processed data and a planetary model. First we suppressed the starlight by a per-night average after Doppler shifting each spectrum to the rest frame of the star. Since the planetary Doppler wobble is of the order of several km s^{-1}, this signal is smeared for a long enough observation run (Fig 1), leaving only the star's signature. This averaged spectrum is then subtracted to each frame, After that only the planetary signature remains submerged in the noise.

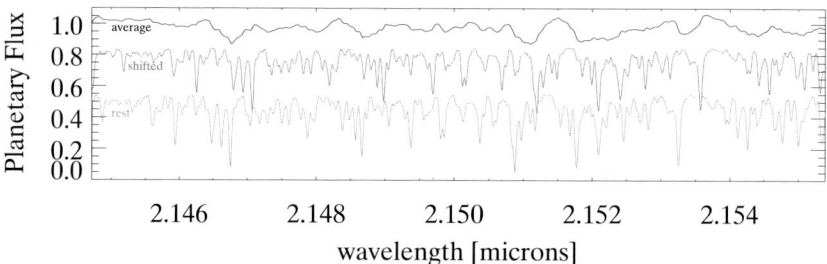

Figure 1. Light-gray line shows a synthetic emission spectrum of HD 217107b at rest. The gray line shows the same as before shifted 40 km s^{-1}, equivalent to the radial velocity span of a single night of observation. the top black line shows the average after considering all the spectra observed between the previous two cases. The planetary spectrum is clearly smeared.

Depending on the unknown inclination i of the orbit, the radial velocity of the planet will be a distinctive curve in time as a function of known orbital parameters. We then cross-correlate the doppler-shifted synthetic spectra (according to an assumed value of $\sin(i)$) with the processed data. By repeating this step for the different values of $\sin(i)$, we should obtain a positive correlation when i matches value of the planetary system (Fig 2). A strong peak in the parameter space of the correlation function represents a successful detection of the planetary signature and indicates the value of the inclination of the orbit.

An important difference with the Wiedemann et al. work is that we constructed an averaged template for each night, as the atmospheric conditions could change greatly from night to night.

3. Results

The innermost planet, HD 217107b, orbits at 0.073 AU from its host star every 7.1 days (Fischer et al. 1999; Butler et al. 2006). A customized planetary model was used for this project, modeling three effective temperatures for this eccentric system. (algorithm described in Fortney et al. (2005)), and convolved down to the instrumental resolving power. On our preliminary results we can only determine an upper limit of the planet-to-star flux ratio of 2×10^{-3} (Fig 3), based in comparisons of the degree of correlation with correlation of simulated data with an injected planet signature. Unfortunately, the instrument was not well characterized at the time of the observations leaving much larger acquisition overheads than expected.

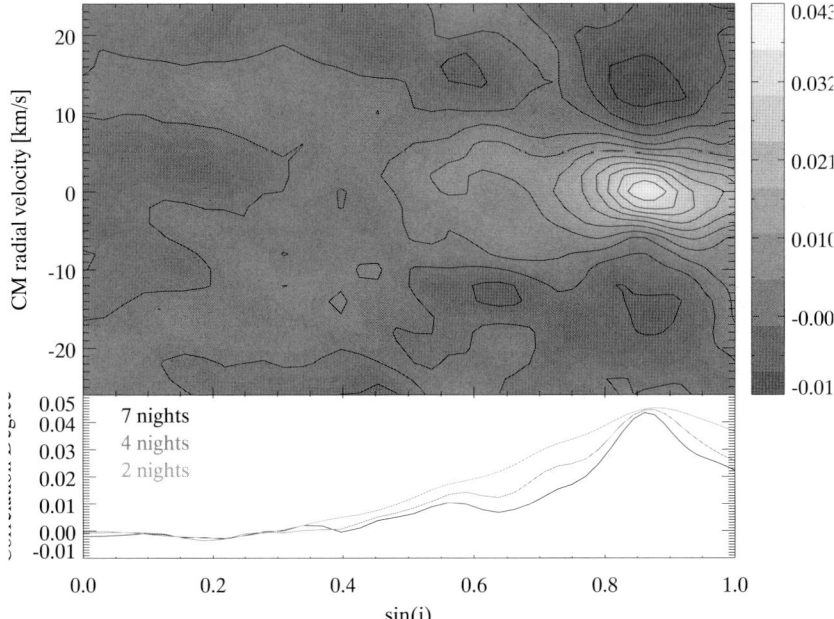

Figure 2. Top: cross-correlation results as a function of sin(i) and radial velocity of the center of mass, v_g, for synthetic CRIRES observations of 7 half-nights (35 hours) of HD 179949b with a planet-to-star flux ratio of 10^{-4} and sin(i) = 0.86, the brightest colors denote positive correlation. The known parameter v_g is used as a variable to test the procedure, since each night should trace a straight line in the map. Bottom: Black line shows a slice of the map at the top at the expected radial velocity of the center of mass of the system as a function of the unknown parameter sin(i). Gray and light-gray lines show a simulation of the resulting correlation with only 20 and 10 hours of observations, respectively. The peak gets more distinctive as more time is involved.

A statistical method to assess confidence levels to claim a detection or set upper limits in the planetary flux is currently under development.

4. Acquisition Strategies and Simulations

In the search of a better candidate, those systems which maximize the success of this technique are those with a close orbiting and short period planet, enabling a higher planet-to-star flux ratio, and a higher radial velocity span per night.

We carried out simulations to measure the effectiveness and sensitivity limit of an observing campaign with a better instrument (CRIRES, R=50000) and target (HD 179949b) We simulated observations according to the orbital parameters, selecting the best nights to cover as much of the orbital phase as possible and to have a high

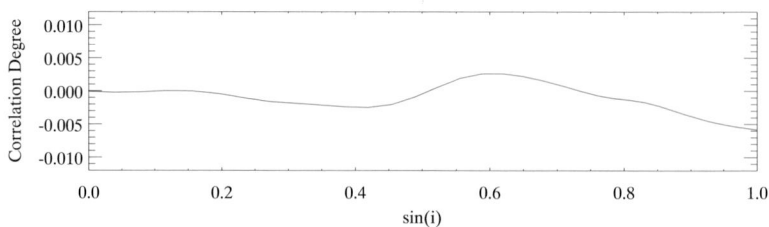

Figure 3. Model-data correlation. The correlation degree found was very low without a remarkable peak.

radial velocity span per night (Fig 2). We used a high-resolution solar spectrum[2] and the synthetic model for HD 217107b to simulate the solar and planet spectrum. Although in this case the model does not correspond system that we are simulating, most important is that we reproduced the planet-to-star flux ratio expected and added the corresponding Poisson noise.

We have found a sensitivity bias, tending to favor the detection of more inclined orbiting systems leaving a dependant sensitivity limit in the planet-to-star flux ratio at each inclination. This is expected as greater inclinations implies less massive planets, therefore, appear more smeared in the averaged spectrum.

5. Conclusions and Future Prospects

Although we could not constrain new parameters of HD 217107b, we have shown that using realistic simulations of the data that this method could yield successful results with the right observational strategy.

As these are preliminary results, remain yet to elaborate a routine to determine the statistical significance of the results, and look for possible systematic errors.

Acknowledgments. We acknowledge the support from Centro de Astrofísica FONDAP 15010003.

References

Butler, R.P., Wright, J.T., Marcy, G.W., Fischer, D.A., Vogt, S.S., Tinney, C.G., Jones, H.R.A., Carter, B.D., Johnson, J.A., McCarthy, C. & Penny, A.J. 2006, ApJ, 646, 505
Fischer, D.A., Marcy, G.W., Butler, R.P., Vogt, S.S. & Apps, K. 1999, PASP, 111, 50
Fortney, J.J., Marley, M.S., Lodders, K., Saumon, D. & Freedman, R. 2005, ApJ, 627, L69
Horne, K. 1986, PASP, 98, 609
Wiedemann, G., Deming, D. & Bjoraker, G. 2001, ApJ, 546, 1068

---

[2]http://bass2000.obspm.fr/solar_spect.php

Detection of Extrasolar Comets

Olivier R. Hainaut

ESO, Karl Schwarzschildstraße 2, 85768 Garching-bei-München, Germany

Abstract. Realistic and extreme cases of Solar System comets have been scaled to the distance of a neighbour star, to investigate the detectability of extrasolar comets. While the nuclei are not detectable, the tail could be observed in some exceptional cases. Transiting comet can cause a detectable extinction, and have already been observed with high resolution spectroscopy.

1. Introduction

The purpose of this short paper is to crudely evaluate the detectability of extrasolar comets. This is done by extrapolating the characteristics of our Solar System comets to an hypothetical host star at 10 pc, but also by extrapolating real comets to extreme cases. More subtle considerations, such as the difference between host star and the Sun, and possible physical differences between our comets and extrasolar ones are not taken into account.

2. Comet Nucleus

Cometary nuclei are small objects. Observing them at inter-stellar distances is virtually hopeless, especially as they would appear at a small angular distance from their home star. This section aims at quantifying this.

A Sun-like star is considered for the illumination of the nucleus. The illumination phase effect is not accounted for; at quadrature (maximum angular separation), the phase effect will decrease the brightness by almost $\sim$ 1mag. A typical comet nucleus has a radius in the $0.1-100$km range (eg 5.2km for 1P/Halley (Keller et al. 1987), 30km for c/1995 O_1 Hale-Bopp (Fernández 2000), a large comet; a TNO falling in the inner Solar System would be in the 100km range. The typical albedo for cometary nucleus is $p \sim 0.05$ (Campins & Fernández 2000; Lamy et al. 2004).

Scaling the light reflected in the visible domain for the distance to the star and to the Earth, the 0.5μm, V band magnitudes of these nuclei are in the 28-45 range (depending on the size of the object and its distance to the star), i.e. orders of magnitude below the limit for direct detection, especially near a bright star, even with the next generation of Adaptive Optics systems. Typical colours for the cometary nuclei are $V-J = 1.6$ and $V-K = 2.1$ (Hainaut & Delsanti 2002), making the situation even worse in the near infrared. Moving to the thermal emission in the infrared, we scaled observations of 73P/Schwassmann-Wachmann 3 and adjusted thermal models of cometary nuclei (Mueller, Priv.Comm.) to estimate the flux of 73P- and Hale-Bopp-like objects at ~ 1 AU from the star. At 10 and 20μm, their flux would be in the 10^{-14}–10^{-13} Jy and 10^{-9}–10^{-8} Jy ranges, respectively, which are again orders of magnitude below what will be detectable in a reasonable future.

For completeness, the star radial velocity change caused by an orbiting comet is of the order of 10^{-8} – 10^{-13} m/s, while the best instruments are now barely entering in the dm/s range.

3. Comet Tail

Let's now consider the case of the comet tail. Halley and Hale-Bopp have fairly large perihelion distances (q = 0.6 and 0.9 AU, resp.), so we first scale them to q = 0.1 AU as a reasonable value (we label these "Super Halley" and "Super Hale-Bopp"), and to q = 0.01 AU as an extreme distance at which the comet is likely to completely evaporate ("Exploding Halley" and "Exploding Hale-Bopp"). The evolution of the total magnitude of a comet with heliocentric distance r varies widely from comet to comet, so the customary scaling law in $2.5 \log(r^{-4})$ was used. They were also scaled for the case of a giant, TNO-like nucleus. These objects were then placed at a geocentric distance of 10 pc. The J band values are obtained from the colours of Halley's tail from Morris & Hanner (1993). The corresponding magnitudes are listed in Table 1.

Table 1. Integrated magnitudes for comets around a Sun-like star at 10 pc

Radius [km]	Comet	q [AU]	V	J
5	Halley	0.6	32	36
5	Super Halley	0.1	24	28
5	Exploding Halley	0.01	14	18
35	Hale-Bopp	1	30	34
35	Super Hale-Bopp	0.1	20	25
35	Exploding Hale-Bopp	0.01	**10**	15
100	TNO-sized super comet	0.1	17	21
100	Exploding TNO	0.01	7	11

While the historical comets observed from 10 pc would be below the detection threshold, a large or very large object passing very close to the star, will develop a tail with a total magnitude in the detectable range. Changing the star from a sun to a K or M-type star, the comet would be 2 or 5 mag fainter (resp.). While the contrast between the comet and the star would remain constant, the absolute flux of the comet is likely to become too faint for detection. The visible domain contains several strong emission lines (see A'Hearn 1982, Fig. 1 and Fig. 5), in particular the C2 bands around at 0.51μm and OI and NH_2 at 0.63μm and the very strong CN at 0.39μm. Observing at these wavelengths will greatly increase the contact between the comet and its star.

In summary, a large to very large cometary nucleus passing very close to the parent star is likely to develop a tail whose integrated flux should be detectable. The flux in comet emission line will not only be higher, but also have a much stronger contrast. The contrast with the star will be challenging, however the main question is the frequency of such event. In our Solar System, they are very infrequent (we had one Hale-Bopp in historical times, and no known TNO-sized comet). Of course, in a younger system, one can expect more objects.

4. Transiting Tail

It is possible to estimate the extinction caused by a comet passing in front of its star (even not knowing the actual dust grain size distribution) by considering the $Af\rho$ measurements of solar comets. Af is the "filling factor" of an image element containing cometary dust (A is the albedo of the grain, f the fraction of the area of the picture element actually covered by dust). In steady state, the profile of a comet drops as $1/\rho$ (ρ is the projected linear distance to the nucleus), consequently the $Af\rho$ factor (which has the dimension of a length) is fairly constant over the extension of the coma (see A'Hearn et al. 1984, for complete description of $Af\rho$). $Af\rho$ has been measured for many bright comets, with typical values around a few hundred metres; the highest value was measured on Hale-Bopp near perihelion (Weiler et al. 2003), $Af\rho = 5$ km.

Considering the coma of a giant comet can be several 10^5 km, a comet transiting in front of the star can "cover" all the diameter of the star if the geometry is appropriate. Using the common $A = 0.04$ for the grains, $Af\rho$ can then be converted into a total extinction of the stellar disk by the dust in the coma. The table below lists the total extinction (Ext, a fraction from 0 to 1), and the radius of the equivalent opaque disk causing the same extinction, r_e.

Table 2. Transit of a comet; r_e is the radius of the equivalent opaque disk.

R_* [R_{sun}]	[km]	Type	r_e [km]	Ext.
0.15	1.1×10^5	M5V	5.1×10^3	2.4×10^{-3}
0.19	1.3×10^5	M4V	5.8×10^3	1.9×10^{-3}
0.3	2.1×10^5	M3V	7.3×10^3	1.2×10^{-3}
0.42	2.9×10^5	M1.5V	8.6×10^3	8.5×10^{-4}
0.75	5.3×10^5		1.2×10^4	4.8×10^{-4}
1	7.0×10^5		1.3×10^4	3.6×10^{-4}

These are the values for an exceptional comet (HB), but at a fairly distant perihelion. Values at least one order of magnitude stronger are expected for a closer perihelion distance and for a larger comet. The central part of the comet, close to the nucleus, has a denser content of dust, and therefore contributes more to the extinction. Therefore, the relative extinction is stronger for small stars than for larger stars. In some cases, projection effects will also bring the tail of the comet in the line of sight, increasing again the extinction. As the comet will be between the observer and the star whose radiation pressure pushes the dust away radially, this configuration is actually likely to happen. The table lists the geometric extinction at wavelengths where the dust grains are opaque. This holds for most of the visible and NIR range. At longer wavelengths, one expects the extinction to be less efficient. The coma and tail containing a large number of absorbing species, it is expected that high resolution spectroscopy will reveal tell-tale absorption patterns. Using this method, extrasolar comets were observed around β Pic (Beust et al. 1990, and references therein), years before the first extrasolar planet discovery, by absorption features in the various metallic lines.

5. Summary

While the nuclei of extrasolar comets are and will remain out of reach for any foreseeable future, direct detection of the coma and tail of exceptional comets might be feasible. This would however require the quasi explosive sublimation of an extremely large nucleus (TNO-sized), which is an very unlikely occurrence in a mature planetary system, but might happen in very young systems. The observations would be slightly less challenging when performed in some of the cometary emission line, where the contrast between the coma and the star will be weaker. Nevertheless, the most promising way to detect extrasolar comets seems to be via transit. The overall absorption by the coma dust can be stronger than 10^{-3} for a large comet, and could be much stronger if the geometry is favourable (ie the tail also contributing to the transit), or in case of an exceptionally active comet. Additionally, specific absorption by sublimation products will produce a much stronger signature, which has already been observed in the case of β Pic.

The case of extrasolar comets is therefore not as hopeless as it could have appeared.

References

A'Hearn, M. F. 1982, in IAU Colloq. 61: Comet Discoveries ..., ed. Wilkening, 433–460
A'Hearn, M. F., et al. 1984, AJ, 89, 579
Beust, H., Vidal-Madjar, A., Ferlet, R., & Lagrange-Henri, A. M. 1990, A&A, 236, 202
Campins, H. & Fernández, Y. 2000, Earth Moon and Planets, 89, 117
Fernández, Y. R. 2000, Earth Moon and Planets, 89, 3
Hainaut, O. R. & Delsanti, A. C. 2002, A&A, 389, 641
Keller, H. U., et al 1987, A&A, 187, 807
Lamy, P. L., et al 2004, Cometary nuclei (Festou, et al), 223–264
Morris, C. S. & Hanner, M. S. 1993, AJ, 105, 1537
Weiler, M., Rauer, H., Knollenberg, J., Jorda, L., & Helbert, J. 2003, A&A, 403, 313

Part VIII

The Reception

The Reception

Location and Social Events

The conference took place in the Cassini Hall of Observatoire de Paris. On November 20, the hosts of the conference had organised a special evening buffet that consisted of a luxury selection of fine wines, canapés, regional french cheeses and charcuterie, round off by an elegant selection of different desserts, typical to France. This evening reception began at 19:30, with a unique dance celebration "COSMOS 1" created by Tamara Millon.

L'Observatoire de Paris

In 1665 the physicist and astronomer Auzout convinced Colbert and Louis XIV to construct 'l'Observatoire Royale'. It is built without wood (to avoid fire) or metal (to avoid magnetic disturbances). At the summer solstice of 1667, the orientation (north-south) was traced in its place by members of the Académie Royale]. Claude Perrault (the architect of the Louvre Colonnade) projected?? the building and directed its construction. It was finished in 1672. It is a large rectangle (31 m × 29 m) with its four faces oriented with the cardinal points of the compass. The latitude of the south face defines the Paris latitude (48°50' 11"). The meridian line passing through its center defines the Paris longitude. The foundations are as deep (27 m) as high is the building itself. The Observatoire is in charge of the French legal time: UTC(OP) and of the Central Bureau of the International Earth Rotation Service. In 1933, the first speaking clock in the world started to give the accurate time by telephone (tel. 3699) from the ground floor of the Observatoire. The basement of the Observatoire is connected with the Paris catacombs (visits forbidden). The catacombs consist of 65 km of underground galleries.

First, at the head of the Observatoire de Paris was Jean-Dominique Cassini (Cassini I), born in Italy in 1625. He was followed by his son Jacques (Cassini II), his grandson César-François and his grand-grand-son Jean-Dominique. The Observatoire was later leaded by Joseph Jerôme Lefrançois de Lalande, Pierre-André Méchain, François Arago (1843-1853), Urbain Le Verrier (1854–1870 and 1873–77) and other distinguished personalities. Further illustrious scientists worked at the Observatoire like Jean-Baptiste Delambre, Charles Marie de La Condamine and Pierre Simon de Laplace.

One can mention some important scientific work conducted at the Observatoire during its long history:

- The map of the Moon by Cassini I that was the best till the photography was invented.

- The discovery of the gap in the Saturn's ring system by Cassini and the table of the satellites of Jupiter movements that allowed the danish astronomer Olaüs Römer to show that the speed of light was finite and compute approximately, for the first time in 1676 while he was working at the Observatoire.

- Jacques Cassini discovered the proper motion of Arcturus, showing for the first time, that the stars were not fixed.

- César-François and Jean-Dominique (IV) Cassini made the first modern map of France from 1750 to 1790.

- The units of mass (gramme) and length (meter) were defined following measurements (along the French meridian) and other research conducted at the Observatoire. Also, Lavoisier worked here on the mass unit.

- Arago introduced photography into astronomy. In 1845 Hyppolite Fizeau and Léon Foucault obtained the first daguerreotype of the Sun.

- Foucault in 1850–51 showed manifestly the rotation of the earth with his pendulum hanging in the Cassini hall (after a first experiment at his home, and before the demonstration at the Panthéon).

- The works by Le Verrier lead to the discovery of Neptune. His tables of solar and planetary positions were used for more than a century. Discrepancies noted by him between the calculated and observed orbit of Mercury were only re-solved with the advent of general relativity.

The three main halls in the Observatoire are the 'Grande Gallerie' and the 'Salle du Conseil' in the ground floor and the 'Salle du Cassini' on the first floor. In the 'Salle du Conseil' are portraits of Laplace, Le Verrier, Lalande, Arago, Delambre, as well as other distinguished scientists and the one of Louis XIV. At present the Observatoire de Paris owns three campuses: Paris, the Meudon astrophysics section and the radioastronomy station at Nançay. More than 700 scientists, technicians and administrative staff work at these three institution. (Courtesy Ecole Challonge Website, Pr Norma Sanchez)

The Reception

Observatoire de Paris was also the stage of the Hollywood block buster "From Paris with Love" starring John Travolta during the 2008 molecule conference.

Artistique creation by Tamara Millon *COSMOS 1*

During the reception Tamara millon presented an artistic creation mixing, music, dance and theater.

Artist statement. Il s'agit d'une nouvelle rencontre avec la/les matières artistiques: la danse, le théâtre et les arts plastiques, que l'on appelle aujourd'hui art visuel. Cette rencontre sort des cadres traditionnels, elle prend la forme d'une performance et nous emmène dans un univers onirique: celui de l'artiste d'origine russe Tamara Millon.

Concept. Donner à voir les oeuvres plastiques, sur des sonorités venues d'ailleurs ainsi que le corps en mouvement, le tout mis en scène pour une vraie rencontre avec le public… (durée de l'œuvre originale 1,15 min) Il s'agit également d'un voyage dans le temps, dans le cosmos, dans l'infini, dans les symboles. Notre monde matériel se confronte à l'autre monde, le cosmos, traduit au travers du vocabulaire plastique et virtuel d'aujourd'hui. L'artiste a créé pour l'occasion une série d'éléments plastiques, de costumes, et d'images.

Le projet COSMOS 1. Comment traduire le lien entre hier, aujourd'hui et demain? ici et ailleurs, face à ce grand Tout? Qui n'a jamais rêvé d'aller dans l'espace? Pour cette nouvelle approche et pour le public, T. MILON dans la continuité de ses recherches en art visuel, a choisit le corps vecteur d'art en mouvement, "électron libre," vitesse, lumière... Chaque Matière Artistique qui apparait, prend corps dans un personnage mis en scène, l'œuvre est en quelque sorte un ensemble de gestes dansés, de déplacements, d'instants évocateurs lumières donnés à voir. INSTANTS-VITESSE humanisés par l' artiste qui puise dans «ce tas de matières»: d'un élément géométrique rappelant l'espace, à l' immensité du geste, une oeuvre virtuelle qui passe... Les toiles devenues vivantes symbolisant ou représentant le ciel, les planètes, jusqu'aux costumes originaux peints expressifs, tels ceux créés au début du 20e siècle par le Bauhaus, ou à l'époque des Ballets Russes, références ou simples images vivantes, à chacun d'y voir. Un voyage vers un ailleurs, une immensité , un voyage intemporel pourtant contemporain. Cette performance, lieu de résonnance sur les traces de ce passé, de ce présent, de ce demain énigmatique et à la rapidité technologique, témoigne de la vie sur terre, dans l'univers, comme l'envolée lyrique créée par la présence féminine et légère de l'artiste Tamara MILON en sculpture vivante-planète ou en étoile filanteĚ

En résumé. De la liberté du mouvement dansé esquissé, jusqu'à la confrontation de l'espace harmonie, de l' espace dérive, les sons invisibles de ces corps histoires, de ces corps vibratoires, de ces corps respirants et traçants la/ les trajectoires humaines, histoire d' une interrogation, d'une autre trajectoire couleur d'infini, sans limite: le cosmos. L'impalpable présence mais pourtant sensible et sensitive vision de cette mixité pluridisciplinaire contemporaine imaginaireĚ Les œuvres, les couleurs, les images, la succession des apparitions, des disparitions, "des gestes vitesses lumiéres," leurs juxtapositions enrichissent cet ailleurs invisible rendu visible vers une élévation onirique et visionnaire.

Tamara Milon, Novembre 2008.

Tamar Milion performance at the reception.

Blind Wine Tasting in Observatoire de Paris

This conference was marking the dawn of complex molecules retrieval in extrasolar planet atmospheres, so it looks to us as a perfect moment in which to attempt further inverse problem solving—blind wine tasting! We set up two tables with high quality wines to taste. On the left side of the table were the Rhone valley wines, and on the right side were the Bordeaux. The idea is to take you on a short journey, through two very distinct regions of France. Wines were only identified by numbers. The participants had a list describing the 10 wines. The game was to figure out which wine was which!

The Bordeaux Wines

- **Château du Pavillon, Sainte Croix du Mont, 1998** (white), 85% Semillon, 15% Sauvignon blanc. A very well made St. Croix du Mont, with golden color, rich and ample flavor.

- **Château Cote de Montpezat, Cote de Castillon, 2001** (red), 70% Merlot, 10% Cabernet Sauvignon, 20% Cabernet Franc. Tanic and well made wine. Beppe–Crosarial (Globe and Mail): "The cote de Montpezat delivers pure cassis like fruits as well as nice, classic Bordeaux notes and mineral and pencil shavings and juicy acid."

- **Château la Parde de Haut Bailly, Pessac Lèeognan, 1999** (red) 56% Cabernet Sauvignon, 28% Merlot, 15% Cabernet Franc. Second wine of chateau Haut Bailly, Pessac Lééognan. Beautifull floral nose, some meaty elements alongside some attractivity perfumed red fuit. Gentle, moderately concentrated, rather firm and intense tanins, light but smooth finish.

- **Château de Haut Marbuzet, Saint Estèphe 2003** (red), 50% Cabernet Sauvignon, 40% Merlot, 10% Cabernet Franc. Complex and elegant, very well balanced, one of the best St Estèphe wines in an excellent year.

Most of the participants recognized that the sweet white wine was the Saint Croix du Mont. Sorting out the three Bordeaux reds was tricky. The Saint Estéephe was quite well identified, but there was a bit of confusion between the Cote de Castillon (small name, but well made), and the Pessac Lèognan.

- Sweet wines from Bordeaux: You immediately think about Sauternes! Here, we provide you with a Sainte Croix du Mont; a less known area, a satellite of Sauternes. Something to keep in mind when choosing your wine. Don't forget the so called satellites! Such wines can age beautifully, gaining in complexity. This one was 10 years old, good to drink now. A well made wine, from a good year in a "satellite" will make you happier than a regular wine from a more famous area!

- Cote de Castillon: It is a typical Bordeaux wine, with Merlot, Cabernet Sauvignon and Cabernet Franc. Only 30 km away from Bordeaux, towards St. Emilion. It shows you that you do not need to have a prestigious name (I guess most of you have not heard of Cote de Castillon before), but when well made, it is very pleasant.

- La Parde de Haut *Bailly is a "second wine," meaning you have a "grand cru classe" the Château Haut Bailly, and the very same people are doing a wine with the same passion, in the same area, but that is not entitled to have the name "Grand Cru Classe" (because the plot of land does not have this label). Therefore, it is sold cheaper, but it is of excellent quality. That's one thing to remember "on good years, do not hesitate to go for the second wines of big names". This is most of the time excellent value !

- Chateau de Haut Marbuzet is a famous Saint Estephe wine (Médoc area), well made, a classic wine.

In France, it is usually considered rude to speak about the prices of the bottle you put on the table. However here, we think here it is a good opportunity to show you that you can get nice wines, for very decent prices. The Saint Croix du Mont is priced at 10 euros, the Cote de Castillon 8 euros, the Pessac Léognan 19 euros, and the Saint Estephe 23 euros.

The Rhone Valley Wines

- **Laurent Combier, Croze Hermitage 2007** (red). It is a 100% Shiraz*??, with typical flavors from the South East of France, blackberry, prune, olive, long finish, smooth and velvety.

- **Le Parvis, Châteauneuf du Pape, 2004** (red), Grenache and Mourvendre grapes. Dark red, color, very intense. Thym, rosmary, fig, pears and spices. Very long finish, warm.

- **La Galopine, Delas Fréères, Condrieu, 2006** (white), 100% Viognier, a grape that provides sensuality, but that could be also "capricious." Here is a superb result. Robert Parker wrote about this one: "Peach apricot, lychee nut characteristic in its stylish fruit forward personnality." Usually with Condrieu, one finds peach, abricot and also violet flavors.

- **Château Mont-Redon, Château Neuf du Pape, 2003** (white), mostly white Grenache, with also Bourboulenc, Clairette and Roussane. Powerful, rich taste, well balanced with spices and white fruits.

The easy one to identify was the Croze Hermitage (young full bodied, very bright color, almost a new world wine!). Since it was the youngest, it was pretty straightforward to pick it up in comparison with a more complex one. Therefore the red Chateau neuf du Pape was also identified without too much difficulty. However, distinguishing between the Condrieu and the white Chateau neuf du Pape, it was hard. Both were very well made! In terms of prices, the Croze Hermitage is priced at 7 euros, the two Chateau neuf 25 euros, and the Condrieu 30 euros. We hope that you enjoyed this small trip in these two regions of France!

In the Bordeaux region about half of the participants identified correctly the Chateau Marbuzet, the other half mistook it for the Pessac Léognan or the cote de Castillon. A large fraction of people identified correctly the St. Croix du Mont as well. On the Rhone valley, it was a bit more chaotic, especially between the whites.

The three winners of the molecules 2008 wine tasting are Nikole Lewis, Adam Burgasser & Daniel Benest. Congratulations!

The reception in Cassini hall.

Keigo Enya toasting to SPICA, while Gautam Vasisht is engaged in enlightening conversations.

Wine tasting! For Monique Gros, Daniel Benest, and Daniel Briot it looks like serious buisness! Whereas Laurence Rothmans is taking it easy.

Ferenc Gonter and Florence Chafiol-Chaumont coming from a distant world to share a glass with astronomers. Giovanna Tinetti, Nicole Allard, and Jean-Philippe Beaulieu having a toast.

250 *Observatoire de Paris*

Tamara Milon. "Moon" 1997. 120 cm × 120 cm.

ASTRONOMICAL SOCIETY OF THE PACIFIC

THE ASTRONOMICAL SOCIETY OF THE PACIFIC is an international, nonprofit, scientific, and educational organization. Some 120 years ago, on a chilly February evening in San Francisco, astronomers from Lick Observatory and members of the Pacific Coast Amateur Photographic Association—fresh from viewing the New Year's Day total solar eclipse of 1889 a little to the north of the city—met to share pictures and experiences. Edward Holden, Lick's first director, complimented the amateurs on their service to science and proposed to continue the good fellowship through the founding of a Society "to advance the Science of Astronomy, and to diffuse information concerning it." The Astronomical Society of the Pacific (ASP) was born.

The ASP's purpose is to increase the understanding and appreciation of astronomy by engaging scientists, educators, enthusiasts, and the public to advance science and science literacy. The ASP has become the largest general astronomy society in the world, with members from over 70 nations.

The ASP's professional astronomer members are a key component of the Society. Their desire to share with the public the rich rewards of their work permits the ASP to act as a bridge, explaining the mysteries of the universe. For these members, the ASP publishes the Publications of the Astronomical Society of the Pacific (PASP), a well-respected monthly scientific journal. In 1988, Dr. Harold McNamara, the PASP editor at the time, founded the ASP Conference Series at Brigham Young University. The ASP Conference Series shares recent developments in astronomy and astrophysics with the professional astronomy community.

To learn how to join the ASP or to make a donation, please visit http://www.astrosociety.org.

ASTRONOMICAL SOCIETY OF THE PACIFIC
MONOGRAPH SERIES
Published by the Astronomical Society of the Pacific

The ASP Monograph series was established in 1995 to publish select reference titles. For electronic versions of ASP Monographs, please see
http://www.aspmonographs.org.

INFRARED ATLAS OF THE ARCTURUS SPECTRUM, 0.9-5.3μm
eds. Kenneth Hinkle, Lloyd Wallace, and William Livingston (1995)
ISBN: 1-886733-04-X, e-book ISBN: 978-1-58381-687-5

**VISIBLE AND NEAR INFRARED ATLAS
OF THE ARCTURUS SPECTRUM 3727-9300Å**
eds. Kenneth Hinkle, Lloyd Wallace, Jeff Valenti, and Dianne Harmer (2000)
ISBN: 1-58381-037-4, e-book ISBN: 978-1-58381-688-2

ULTRAVIOLET ATLAS OF THE ARCTURUS SPECTRUM 1150-3800Å
eds. Kenneth Hinkle, Lloyd Wallace, Jeff Valenti, and Thomas Ayres (2005)
ISBN: 1-58381-204-0, e-book ISBN: 978-1-58381-689-9

**HANDBOOK OF STAR FORMING REGIONS: VOLUME I
THE NORTHERN SKY**
ed. Bo Reipurth (2008)
ISBN: 978-1-58381-670-7, e-book ISBN: 978-1-58381-677-6

**HANDBOOK OF STAR FORMING REGIONS: VOLUME II
THE SOUTHERN SKY**
ed. Bo Reipurth (2008)
ISBN: 978-1-58381-671-4, e-book ISBN: 978-1-58381-678-3

A complete list and electronic versions of ASPCS volumes may be found at
http://www.aspbooks.org.

All book orders or inquiries concerning the ASP Conference Series, ASP Monographs, or International Astronomical Union Volumes published by the ASP should be directed to:

Astronomical Society of the Pacific
390 Ashton Avenue
San Francisco, CA 94112-1722 USA
Phone: 800-335-2624 (within the USA)
Phone: 415-337-2126
Fax: 415-337-5205
Email: service@astrosociety.org

For a complete list of ASP publications, please visit
http://www.astrosociety.org.